中国工程科技论坛——黏菌分类学及生态学研究部分与会代表合影 (2014.8.12)

中国工程科技论坛

黏菌分类学及生态学研究

nianjun fenleixue ji shengtaixue yanjiu

Taxonomy and Ecology of Myxomycetes

高等教育出版社·北京

内容提要

黏菌是介于真菌和植物之间的一类特殊生物，现被归在原生动物界。为了深入探究黏菌的奥秘及展望今后的科研动态，2014年8月在吉林长春举办了第186场中国工程科技论坛——黏菌分类学及生态学研究，与会嘉宾分别来自美国、英国、俄罗斯、德国、中国、日本、意大利、瑞士、法国、比利时、印度、乌克兰、菲律宾，这些顶尖专家们围绕"黏菌学——一个极具挑战性的领域"的主题分别进行了主题演讲和学术讨论。本书是在对这些论坛报告进行汇编整理的基础上而成，内容丰富，涵盖了黏菌的生态学、生物地理学、生物多样性、分类学、分子系统学及生物学等领域的学术论文及各位专家的报告内容。

本书是中国工程院"中国工程科技论坛"丛书之一，是一本有重要参考价值的专著，可供生物学、菌物学和药学研究者参阅。

图书在版编目(CIP)数据

黏菌分类学及生态学研究 / 中国工程院编著. —— 北京：高等教育出版社，2015.8
（中国工程科技论坛）
ISBN 978－7－04－043025－7

Ⅰ. ①黏… Ⅱ. ①中… Ⅲ. ①真细菌目－研究 Ⅳ. ①Q939.11

中国版本图书馆 CIP 数据核字（2015）第 128539 号

总策划　樊代明
策划编辑　王国祥　黄慧靖　　责任编辑　朱丽虹
封面设计　顾　斌　　　　　　责任印制　张泽业

出版发行	高等教育出版社	咨询电话	400－810－0598
社　　址	北京市西城区德外大街4号	网　　址	http://www.hep.edu.cn
邮政编码	100120		http://www.hep.com.cn
印　　刷	北京丰源印刷厂	网上订购	http://www.landraco.com
开　　本	787mm×1092mm		http://www.landraco.com.cn
印　　张	12.75		
字　　数	230千字	版　次	2015年8月第1版
插　　页	1	印　次	2015年8月第1次印刷
购书热线	010－58581118	定　价	60.00元

本书如有缺页、倒页、脱页等质量问题，请到所购图书销售部门联系调换
版权所有　侵权必究
物　料　号　43025－00

编辑委员会

主　编：李　玉
副主编：王　琦　刘淑艳　刘　朴
编　委：蒋世翠　安　颖　王欣欣　陈梁城

目 录
Table of Contents

第一部分 综述
Part I Review

综述 ———————————————————————— 李玉　3

第二部分 主旨报告
Part II Keynote Speech

More Than 50 Years with Myxomycetes (Plasmodial Slime Molds): Highlights and Review ———————————————————— Harold W. Keller　7

Myxomycetology: A Challenging and Inspirational Field ———— Yu Li　27

Resource Allocation and Morphogenesis during Fructification in Myxomycetes
———————————————————— Indira Kalyanasundaram　35

中国黏菌生物学研究进展 ———————————————— 王琦, 李姝　46

第三部分 生态学、生物地理学以及生物多样性
Part III Ecology, Biogeography, and Biodiversity

Myxomycetes and Protosteloid Amoebae in the Man and Biosphere Reserve at Yangambi (D. R. Congo) ———————————— Myriam De Haan, Christine Cocquyt, George G. Ndiritu　59

Comparative Diversity of Myxomycetes in Paleotropical (Philippines) and Temperate (USA) Forests ———————————— Thomas Edison E. Dela Cruz, Adam W. Rollins, Steven L. Stephenson　61

Myxomycetes in Forest Patches on Ultramafic and Volcanic Soils: Assessment of Species Diversity and Heavy Metal Biosorption
———— Maria Angelica D. Rea, Nikki Heherson A. Dagamac, Fahrul Zaman Huyop, Roswanira A. B. Wahab, Thomas Edison E. Dela Cruz　63

Looking at the Diversity of Myxomycetes in the Limestone Forests of Puerto Princesa Subterranean River National Park in Palawan, Southern Philippines ———— Melissa H. Pecundo, Thomas Edison E. Dela Cruz　65

More Additions to the Checklist of African Myxomycetes
———— George G. Ndiritu, Myriam De Haan　67

Digitalization of the Types from the N. E. Nannenga-Bremekamp Myxomycetes Collection xx ———— Myriam De Haan, Ann Bogaerts　69

Myxomycetes Growing on Epiphytic Bryophytes: An Opportunity
———— Myriam De Haan　70

Some Ecological Aspects of Nivicolous Myxomycetes of the Khibiny Mts. (Kola Peninsula, Russia) ———— D. A. Erastova, Yu. K. Novozhilov, M. Schnittler　71

Nivicolous Species of *Diderma* spp.: Morphology vs. Genetics
———— D. A. Erastova, Yu. K. Novozhilov, M. Schnittler　73

Nivicolous Myxomycetes in Agar Culture: First Results and Remaining Problems ———— O. N. Shepin, Yu. K. Novozhilov, M. Schnittler　74

Passportication for Myxomycetes Conservation ———— Tetiana Kryvomaz　75

Myxomycetes Diversity in Ukrainian Forests ———— Tetiana Kryvomaz　77

Higher Myxomycete Diversity in Mountainous Vegetation than Agricultural Plantation? —An Evidence from Mt. Kanlaon National Park, Negros Occidental, Philippines ———— Julius Raynard Alfaro, Donn Lorenz Alcayde, Joel Agbulos, Nikki Heherson Dagamac, Thomas Edison E. Dela Cruz　84

A Look at the Diversity of Myxomycetes in the Mountain and Coastal Forests of Puerto Galera, Oriental Mindoro ———— Nathan S. Batungbacal, Carmela Rina T. Bulang, Akira Gioia R. Cayago, Soohyun Jung, Nikki Heherson A. Dagamac, Thomas Edison E. Dela Cruz　86

Myxomycete Diversity and Ecology in Tropical Forests of Southern Vietnam: First Results and Perspectives ——————— Yu. K. Novozhilov, Yu. A. Morozova, A. V. Alexandrova, E. S. Popov, A. N. Kuznetzov 88

Four Years in the Caucasus: Observations on the Ecology of Nivicolous Myxomycetes ——————— Martin Schnittler, Daria A. Erastova, Oleg N. Shchepin, Eva Heinrich, Yuri K. Novozhilov 90

Myxomycetes of Vyatka River Valley ——————— V. A. Sysuev, A. A. Shirokikh, I. G. Shirokikh 92

Myxomycete Diversity and Distribution in the Mountain Valley of Kamikochi in the Northern Japan Alps ——————— Kazunari Takahashi, Yuichi Harakon 99

西藏地区团毛菌目黏菌 ——————— 李姝, 王琦 101

First Report of Sporangia of Two Myxomycetes (*Stemonaria longa*, *Stemonitis splendens*) Collected from Shiitake Cultivation ——————— Bo Zhang, Shicui Jiang, Yu Li 108

第四部分 分类学

Part IV Taxonomy and Systematics

Myxomycetes of Mahe Island in the Seychelles ——————— Tetyana Kryvomaz, Alain Michaud, Steven Stephenson 115

Quantitative Taxonomy? —An Approach for Automated Analysis of Spore Ornamentation from SEM Images ——————— Martin Schnittler, Anna Ronikier, Paulina Janik, Yuri K. Novozhilov 117

Taxonomy, Phylogeny, and Morphological Evolution of the *Polysphondylium pallidum-P. album* Complex (Dictyosteliomycetes) ——————— Shinichi Kawakami 119

Dictyostelids from Jilin Province, China ——————— Pu Liu, Yu Li 121

Dictydiaethalium dictyosporangium sp. Nov. from China ——— Bo Zhang, Yu Li 123

A New Record Species of *Polysphondylium* from China ——————— Mingjun Zhao, Pu Liu, Ying An, Dan Li, Yu Li 124

Revision of the North American *Lamproderma* (Myxomycetes) Collections from the Donald T. Kowalski's Herbarium ——————— Anna Ronikier 125

第五部分 种系发生和遗传学

Part V Phylogeny and Genetics

Comparisons of Genomic DNA Extraction Methods in Myxomycetes —— Pu Liu, Qi Wang, Yu Li ... 129

Molecular Phylogeny of Some Myxomycetes Taxa —— Shuyan Liu, Fenyun Zhao, Yu Li ... 135

The Phylogeny of Slime Moulds (Mycetozoa): from One Gene to the Whole Genome —— Cong Fu, Yu Li ... 137

盘基网柄菌肌动蛋白保守基序的生物信息学分析 —— 李广, 刘淑艳, 李玉, 陈艳秋 ... 138

Amplification and Sequencing of EF-1α Region from *Didymium Squamulosum* —— Shuyan Liu, Fengyun Zhao, Yu Li ... 151

Nuclear DNA Contents of Four Orders of Myxomycetes Collected in Jilin, China —— Shu Li, Bao Qi, Wan Wang, Makoto Kakishima, Qi Wang, Yu Li ... 156

Resource Allocation and Morphogenesis during Fructification in Myxomycetes —— Qian Li, Shuzhen Yan, Shuanglin Chen ... 158

What An Intron May Tell: An Analysis of Two Markers in *Meriderma* spp. (Stemonitales) —— Martin Schnittler, Eva Heinrich, Alexander Kettler, Thomas Sura, Yuri K. Novozhilov ... 160

The Genus *Alwisia* (Myxomycetes) Revalidated, with Three Species New to Science —— Dmitry Leontyev, Martin Schnittler, Steven L. Stephenson, Gabriel Moreno, David W. Mitchell, Carlos Rojas ... 162

New Insights into the *Tubifera ferruginosa*-Complex —— Dmitry Leontyev, Martin Schnittler, Steven L. Stephenson ... 164

第六部分 生物学

Part VI Biology

Application of 3D Imaging of Light and Electron Microscopy in Studying Myxomycetes —— Yuka Yajima ... 169

一种准确测定黏菌原质团原生质流流速的方法 ——————— 王晓丽, 李晨, 李艳双, 李玉　170

A Comparative Study on the Developmental Processes of the Family Physaridae in the Pure Culture ——————— Wei Tao, Shuzhen Yan, Shuanglin Chen　177

Some Hypotheses about Lepidoderma ——————— Renato Cainelli　179

Distribution and the Food Resource Preference of Protostelids in Sugadaira Highlands, Nagano, Japan ——— Y. Iwamoto, Y. Degawa, J. Matsumoto　180

Study on Isozyme in Different Ontogenetic Stages of *Didymium iridis* ——————— Shicui Jiang, Bo Zhang, Yu Li　182

Fatty Acids Detection and Its Application in Taxonomy of Six Dictyostelid Cellular Slime Molds ——————— Ying An, Pu Liu, Yu Li　183

Foraging Behaviors of Phaneroplasmodia in Six Species of Myxomycetes to Three Types of Food Sources ——————— Xiaoxia Song, Bao Qi, He Zhu, Qi Wang, Yu Li　184

Ultrastructure Observations on the Sporulation of *Physarum compressum* ——————— Yanshuang Li, Xiaoli Wang, Yu Li　185

A Preliminary GC-MS Study of Four Species of Physarales ——— He Zhu, Qi Wang　186

Nuclear Observations of *Physarummelleum* ——————— Qi Wang, Shu Li, Yu Li, Makoto Kakishima　188

Liposoluble Constituents Comparison from Five Species of Myxomycetes ——————— Wan Wang, Shu Li, Qi Wang　189

Description of the Amoeboid Movement of Myxamoebae in Several Myxomycetes Species ——————— Xiaoli Wang, Chen Li, Yu Li　190

Species Diversity of Myxomycetes on Different Decay Stages of Coarse Woody Debris in Laurel Forest of Warm Temperate Western Japan ——————— Yuichi Harakon, Shoji Ohga　191

后记 ……………………………………………………………………　193

第一部分
Part I
综 述
Review

综　　述

李玉
中国工程院院士

一、会议的基本情况

2014年8月12日，由中国工程院和中国菌物学会主办，吉林农业大学和吉林省科学技术协会承办，长春科技学院、江苏安惠生物科技有限公司、成都榕珍菌业有限公司协办的第八届国际黏菌系统学及生态学会议暨中国工程院第186场中国工程科技论坛在吉林农业大学和吉林省御龙温泉度假村召开。来自美国、俄罗斯、日本、意大利、印度、乌克兰等15个国家和地区的近60名黏菌学专家和嘉宾共计160人参加了会议。

第八届国际黏菌系统学及生态学会议开幕式在吉林农业大学同声传译报告厅举行。会议由第八届国际黏菌系统学及生态学会议组委会主席、中国工程院李玉院士主持。与会代表首先欣赏了"成长的足迹——历届国际黏菌系统学及生态学会议回顾"视频短片。吉林省政协副主席支建华代表吉林省向与会的各方代表表示欢迎，并宣布第八届国际黏菌系统学及生态学会议暨中国工程院第186场中国工程科技论坛开幕。吉林农业大学校长秦贵信，吉林省科学技术协会主席、中国科学院冯守华院士，中国工程院副院长刘旭院士，吉林省教育厅副厅长苏忠民分别在开幕式上致辞，对大会的胜利召开表示祝贺，并希望借助这一国际学术交流平台，推动我国黏菌学的科技发展，促进国际合作和人才交流。期间Harold W. Keller教授特意从美国发来贺电，对大会的胜利召开表示祝贺。开幕式结束后，与会领导及参会代表集体合影留念，参观了吉林农业大学校园，并对食药用菌教育部工程研究中心、菌菜基地进行了实地考察交流。

出席开幕式的领导和嘉宾还有：中国菌物学会理事长王成树教授，中国工程院二局局长高中琪，长春科技学院院长宗占国，中国工程院夏咸柱院士、李坚院士，中国科学院任露泉院士、魏江春院士，国际药用菌学会执行主席、江苏安惠生物科技有限公司董事长安惠先生，中国科学院微生物研究所菌物标本馆馆长姚一建研究员等。

本次会议分为主旨演讲和学术讨论两部分，围绕黏菌学中最重要的学术议题展开研讨,内容包括:生态学、生物地理学、生物多样性、分类学及系统学、系统发育及遗传学等诸多领域。在主旨演讲环节,国际著名真菌学家、《真菌辞典》主编、英国科学家 Kirk 教授,Harold W. Keller 教授等 5 位学者以"一个极具挑战性的领域:黏菌学"为主题,围绕黏菌分类学及生态学研究,聚焦于黏菌遗传、生物信息、分类等热点问题做了主旨演讲。李玉院士以"一个极具挑战性的领域:黏菌学"为题做了论坛总结发言,并从黏菌认识到利用等方面做了阐述,首次向与会者提出了黏菌学科这一概念,得到与会者的积极响应。在学术讨论环节,与会专家从不同角度,开展多方位、多角度的学术性与前瞻性主题研讨,力求通过多学科的交叉与融合探求世界黏菌发展趋势,加速黏菌科技成果的市场转化,推进世界黏菌研究事业的发展与升级,引领黏菌研究向更专业、广泛和一体化的高度发展。本次会议收到论文 53 篇,27 人做了报告,13 份论文海报进行了展出,还设立了学生组科技论坛,进行了科学考察,举行了黏菌摄影展。

二、专家发言及研讨内容

通过专家发言、讨论,大家一致认为,黏菌是一种神奇的生物,是介于动物和植物间的一类原生动物,在整个生物链上扮演着重要角色。它们点缀着自然界的美丽,为维护生态的多样性贡献着力量。近年来,随着科学技术的不断进步,分子生物信息技术的应用,黏菌在生物计算机的研制等方面的功能与作用,日益引起了科学家的广泛关注。自 19 世纪德国菌物学家德巴里开启黏菌学科的划时代研究开始,在二三百年的时间里,科学家们对黏菌的研究有了系统的发展,取得了辉煌成果,人类对黏菌的认识进入理性阶段。但是,目前世界对黏菌的研究相比于其他微生物还处于较低水平,对黏菌的认识与利用将是一个漫长的过程。

本次论坛主题是"一个极具挑战性的领域:黏菌学",聚焦于黏菌的系统学和生态学研究,这在当前面临资源约束趋紧、环境污染严重、生态系统退化的严峻形势,我国正大力推进生态文明建设的背景下,具有重要的战略意义。在论坛的学术交流中,我们很荣幸地邀请到了来自世界各地知名的黏菌学家,他们从黏菌的分类学、系统学、生态学、生物学、遗传学等各个层面、不同角度去阐述黏菌研究的最新创新成果,使本次论坛成为学术界、产业界、管理层等社会各界人士认识黏菌、了解黏菌、利用黏菌的一次探索之旅,充分进行思想交换和智慧碰撞的学术之旅,提携和培养青年科技人才的攻坚之旅,必定为我国黏菌工程界自主创新的组织管理和自主创新能力的提升起到积极的推动作用。

本次论坛为黏菌研究者搭建了一个高水平的交流平台,报告专家对全球黏菌学科研及产业领域的深刻洞察力及宝贵经验,极大开拓了各界参会代表的视野及思路,对中国黏菌产业发展将产生重大影响。

第二部分
Part II
主旨报告
Keynote Speech

More Than 50 Years with Myxomycetes (Plasmodial Slime Molds): Highlights and Review

Harold W. Keller

Botanical Research Institute of Texas, Fort Worth, Texas 76107, USA

Abstract: My first myxomycete collection of *Dictydium cancellatum* is described. Myxomycete morphospecies concepts are discussed, reference sources given, and criteria and options suggested for the recognition of species new to science. Taxonomic assessment of fruiting body variation is given for *Fuligo septica* and spore ornamentation for *F. megaspora*. Variations of fruiting body characters are discussed for spore-to-spore agar cultures of *Badhamia rhytidosperma* and *B. spinisporum*. A suggested protocol for best taxonomic practice is provided that recognizes the impact of environmental parameters on the plasticity of fruiting body characters using *Cribraria intricata* and *Badhamia rugulosa* as examples. The importance of type collections is discussed, using fine ultrastructure scanning electron microscopy and *Badhamia ovispora* as an example. Monographic publications are emphasized with examples that include *Perichaena brevifila* and *P. reticulospora*. Recent publications documenting spore-to-spore agar cultures with assessment of spore ornamentation and commentary on clustered versus free spores are described.

Key words: Best taxonomic practice; *Badhamia ovispora*; *B. rhytidosperma*; *B. rugulosa*; *B. spinisporum*; *Cribraria intricata*; *Fuligo septica*; *F. megaspora*; moist chamber cultures; morphospecies concepts; *Perichaena brevifila*; *P. reticulospora*; species new to science; spore-to-spore agar culture

1 My fascination with Myxomycetes

My first encounter with Myxomycetes was on a field trip into the mixed hardwood forests near Lawrence, Kansas, while a Master's Degree graduate student in the Department of Botany at the University of Kansas in the early 1960s. Like most beginner collectors I followed the habitat descriptions in books that directed me to decaying logs or leaf litter as the most productive ground sites.

I discovered and collected stalked sporangia of *Dictydium cancellatum* that covered an extensive area on a well-decayed log as part of the ground litter. Upon closer microscopic examination the spore case resembled a bird cage with dark longitudinal ribs connected by hyaline transverse filaments. There was a ball-like mass of aggregated spores freely suspended inside the spore case that was gradually released as the twisted stalk created movements caused by the slightest air currents or physical touch. This salt-shaker method of spore dispersal is often exhibited by moss capsules, for example, *Atrichum* and *Polytrichum*, which release a few spores over time favoring wind dispersal. The myxomycete genus *Cribraria* also often has weaker flexuous stalks that are twisted and facilitates the spores sifting through the peridial network, hence the name *Cribraria* translated from *cribrum*, meaning a sieve.

What fascinated me the most was the intricate structure of the sporangium that reminded me of the phrase "the biological jewels of nature" and the incredible beauty of the form and function represented by myxomycete fruiting bodies. The display of the iridescent brilliant coloration of the *Diachea* or *Lamproderma* species; the intricate ornamentation of the *Hemitrichia* and *Trichia* capillitial threads; the pattern of spore ornamentation from warted to spiny to bordered reticulate, and also clustered spores; the infinite variety of sizes and shapes of fruiting bodies—all make the study of this group of organisms aesthetically pleasing and challenging to find the words that describe this beauty! Yes, beauty is in the eye of the beholder, but for me, the Myxomycetes have piqued my curiosity for more than 50 years.

2 ICSEM 2 Inaugural Lecture

I was invited to deliver the Inaugural Lecture at the 2nd International Congress on the Systematics and Ecology of Myxomycetes (ICSEM 2) held at the Real Jardín Botánico, Madrid, Spain. My paper was published as part of the proceedings under the title "Biosystematics of Myxomycetes: A Futuristic View" (Keller 1996). There were a number of topics of historical interest highlighted such as information relating to myxomycete collectors and mentors Dr. Travis E. Brooks and Professor George W. Martin (the latter my doctoral dissertation supervisor at the University of Iowa), field collecting associates, importance of myxomycete biosystematics, promotion of professional and public interest, best taxonomic practice, importance of ecological field observations, importance of collecting, importance of type collections, importance of spore-to-spore cultivation, living cultures a biological standard, importance of monographic works, importance of computerization of mycological collections, importance of DNA sequencing techniques, future directions, and concluding remarks. I highly recommend that beginning students of Myxomycetes read this paper to better understand some of the guiding principles behind myxomycete systematics.

3 Morphospecies concepts in myxomycete taxonomy

There are five other papers that merit special consideration when applying the morphospecies concept to species new to science: (1) "The Species Problem in the Myxomycetes" (Clark 2000); (2) "Species Diversity in Myxomycetes Based on the Morphological Species Concept — a Critical Examination" (Schnittler & Mitchell 2000); (3) "From Morphological to Molecular: Studies of Myxomycetes since the Publication of the Martin and Alexopoulos (1969) Monograph" (Stephenson 2011); (4) "Myxomycete History and Taxonomy: Highlights from the Past, Present, and Future" (Keller 2012); and (5) "Sporophore Morphology And Development in the Myxomycetes: a review" (Clark & Haskins 2014). These papers discuss some of the problems in myxomycete taxonomy posed by the variability of fruiting body types (sporangium, plasmodiocarp, pseudoaethalium, aethalium) and the basic morphological characteristics used in species descriptions that include but are not limited to (1) fruiting body general habit, color, size of overall dimensions, sessile and stalked in the same collection; (2) peridium when present exhibiting dehiscence patterns, thickness, number of layers, amount of calcareous deposition with either granules or crystals, portion forming a network or cup, inner surface markings; (3) columella presence or absence; (4) capillitium presence or absence, amount and branching patterns and attachments, degree of calcification, capillitial threads with cogs, rings, reticulations or elaters smooth or marked with spines or spiral bands, diameter of width; (5) spore color in mass or by transmitted light, ornamentation either smooth, warted, spinulose, spiny, special markings, bordered reticulate, wall thickness, and spores clustered or free, shape, and size in diameter.

Assessment, evaluation, and examples found in species descriptions in books and journal papers will serve to illustrate potential variation of fruiting body characters. *Fuligo septica* is a myxomycete genus and species characterized by an aethalium fruiting body type and was a species described by Linnaeus in 1780. The myxomycete world monograph by Martin & Alexopoulos (1969) lists *F. septica* with 36 synonyms, including 6 different genera. In the notes following the species description they state:

> "One of the commonest and widely distributed of Myxomycetes. Its extraordinary variability in size, shape, and color is reflected in the numerous names which it has received. The cortex may be very thick or sparse or even lacking, in which case the fruitings appear to be densely clustered and anastomosing sporangia on a common hypothallus, but all have the minutely warted, rather pale spores and there seems to be no way to separate them into coherent subgroups."

Furthermore, the varieties *candida*, *violacea*, *flava*, and *rufa* apparently "do no more than name the color involved" (Martin & Alexopoulos 1969). The color variants are very different in appearance in their striking colors, but an additional suite of distinguishing characters was lacking. However, they also emphasized the following:

> "Unless cultural studies can demonstrate that some of those variations are due to more than a response to conditions under which the plasmodium developed and fruited, they should be disregarded."

Spores of *Fuligo septica* germinate readily in a short period of time, usually within 30 to 90 min (Braun 1971; Keller & Everhart 2010). Fresh aethalia several weeks old gave a higher percentage of spore germination by the split method, forming myxoamoebae and swimming swarm cells as long as free water was present. My students cultured many isolates of *F. septica* from spore to spore on 2% water agar. Yellow phaneroplasmodia fed sterile old fashioned oat flakes were observed and eventually aethalia formed typical of this species. Spores were observed for ornamentation, and at least 100 spore diameters were measured and compared to field collections and found to agree with the species description in Martin & Alexopoulos

(1969): "spores spherical, dull black in mass, purplish, brown by transmitted light, minutely spinulose, 6 – 9 μm in diameter." In other words the spore size and ornamentation were stable in culture suggesting that spore characters were reliable for *F. septica*. In the varieties of *Fuligo* discussed here and in the majority of species (4 of 5) recognized by Martin & Alexopoulos (1969), spores were minutely warted and 6 – 14 μm in diameter. Please note the discrepancy between the species spore ornamentation description and the commentary note that followed (Martin & Alexopoulos 1969).

There was one exception, *Fuligo megaspora*, which merits special consideration. This species was considered rare until Dr. Jean Schoknecht made numerous collections in the Everglades National Park in the state of Florida, USA (Keller & Schoknecht 1989). The spore ornamentation previously was misdescribed using light microscopy (Sturgis 1913; Macbride 1922; Martin & Alexopoulos 1969) as tuberculate arranged in irregular lines, uniformly warted with patches of a close spinulose reticulation, and rough tuberculate to subreticulate markings with the size range of 20 – 22 μm in diameter. This spore ranks as one of the largest in the Myxomycetes and was clearly a *Fuligo* based on the aethalium. This example points out the importance of more accurate species descriptions using scanning electron microscopy that increases the magnification and shows fine structure, in this case, an episporic reticulum with a serrated upper edge (see Figs. 13 and 14 in Keller & Schoknecht 1989). This spore ornamentation is unique in the Myxomycetes. Furthermore, new characters not previously used or recognized sometimes may be helpful in distinguishing species. Granular calcium carbonate ("lime") in the aethalial cortex of *Fuligo* species may show size differences (see Figs. 6 – 10 in Keller & Schoknecht 1989), but this must be tested and compared in different taxa. Spherical calcareous granules ("lime") in the aethalial cortex of *F. cinerea* measure approximately 1 – 1.5 μm in diameter in contrast to that of *F. megaspora* which is more than twice as large at 2 – 4 μm in diameter (Keller & Schoknecht 1989).

Negative culture results often go unreported. Thus, at least 20 mass spore cultures of different *F. megaspora* collections on 2% water agar failed to show spore germination, plasmodial formation, or fruiting body formation (unpublished observations). These starter field collections were fresh (only several months old) but failed to produce spore germination, unlike *F. septica*. Culture work is sometimes frustrating, time consuming, and unrewarding, but more of this kind of research is needed to evaluate species concepts and morphology.

Another surprising discovery examining *Fuligo* collections was a specimen deposited at the US National Fungal Collections (BPI) misidentified as *F. megaspora*, but differed in having spores with long spines (see Figs. 11 and 12, Keller & Schoknecht 1989). The conspicuous long spine was different spore ornamentation from any species of *Fuligo*, and the size range in spore diameter of 16 – 19 μm, was much larger than other *Fuligo* species and more in the size range of *F. megaspora*. Did this taxon represent a new species? More collections were needed, but the collection site of Lignumvitae Key Botanical State Park is an isolated island off the coast of the Florida Keys, USA. that can only be reached via water passage. Dr. Schoknecht was able to visit this island several years later and collected a *Fuligo* along the trail. Upon returning to the laboratory, microscopic examination revealed the long spiny spores and a spore diameter of 16 – 19 μm. Scanning electron microscopic (SEM) images highlighted the long spines not seen in any other *Fuligo* species. Unfortunately the collections and SEM negatives were lost, and attempts to find more specimens failed. This taxon to the best of my knowledge has not been described as new to science in the literature, but it would not surprise me that it will be found someday on that remote island or somewhere else in subtropical habitats. The question still remains, however, whether the spiny-spore character alone is enough to describe a new species or whether associated characters will serve to distinguish this taxon. The next generation of myxomycologists hopefully will search for and discover this taxon and evaluate additional morphological characters.

4 Assessment of variation in spore-to-spore agar cultures

Spore to spore cultivation of two *Badhamia* species, *B. rhytidosperma* and *B. spinisporum*, was also instructive in the variable calcium carbonate deposition in the peridium, in the capillitial system, and also in the formation of a calcareous pseudocolumella (Keller & Schoknecht 1989a, 1989b). Both of these species have unique spore ornamentation: *B. spinisporum* with spores spiny on one side and smooth on the other side and *B. rhytidosperma* wrinkled-reticulate on one side and smooth on the other side with a ridge line marking the future site of spore germination. *Badhamia* and *Physarum* are artificial genera that merge into one another, and many species have been transferred back and forth between the two genera based on the amount of calcium carbonate deposition in the capillitial system. *Physarum* is characterized mostly by a system of hyaline capillitial threads lacking calcium carbonate except for the capillitial nodes, and *Badhamia* is characterized mostly by a network of calcareous capillitial tubules. These two taxa were grown on dung and water agar, and fruiting bodies were compared. Spore ornamentation and size were constant and stable under all conditions of cultivation. Dung cultures had greater concentrations of calcium carbonate in capillital tubules and a more prominent calcareous-filled pseudocolumella when compared to the scanty calcium in agar-developed fruitings (Keller & Schoknecht 1989b).

Premature drying also interferes with normal spore cleavage and development, resulting in aberrant giant spores, and free water or excessive wet surfaces also result in fruiting bodies that lack calcium carbonate deposits and altered shapes (Tamayama & Keller 2013). These same conditions may occur with natural fruitings in the field on different substrata and may give similar results as observed in agar cultures. The best example was *Physarum crateriforme* that had developed on the bark surface of living trees (see Keller & Braun 1999: 129), and the following quotation:

> "Whenever extensive calcareous deposits are present, the white peridium, the columnar columella reaching the upper sporangial apex, and the badhamioid capillitium, coupled with the cylindrical to obovate sporangial shape and black stalk, make this species easy to identify in the field. Under certain conditions it can be troublesome to identify because of extreme variability in sporangial shape and degree of calcification. For example, in some cases the capillitium may consist of hyaline threads lacking calcified nodes and also no apparent columella. These noncalcareous sporangia are often brownish, in contrast to the white color when calcium is present. These fruitings are impossible to identify because spore size and ornamentation fall within the range of many other species."

Persistent long-lasting aethalia of *Fuligo* on ground sites and stalked sporangia of *P. crateriforme* on the bark surface of living trees may undergo physical changes, especially the loss of calcium carbonate deposits due to weathering effects. These morphological changes must be considered by making many collections of the same taxon in the field over time on different substrata to circumscribe the full variation exhibited.

5 Protocol for best taxonomic practice

There is no substitute for collecting specimens in the field and recording the careful observation of habitats, substrata, seasonal occurrence, and variation of fruiting body characters that develop under different environmental conditions. Field collections made from different geographical locations over longer periods of time, including the observation of plasmodia and the formation of the fruiting bodies on natural substrata, result in more accurate species descriptions for monographic works and species descriptions new to science. Single field collections made on single field forays often result in limited material for type specimens and inadequate species descriptions. The following quotation captures the essence of multiple collections made in multiple locations (Keller 2012).

> "One fact should be stressed. The plasmodium may develop its characteristic fruiting stage in less than 24 h. If this occurs under conditions which cause unduly rapid drying or if repeated rains check the process, great variation may be induced. Under such influences, species which ordinarily have stalks may be sessile or nearly so, or the stalks may be inordinately long; sporangiate species may form plasmodiocarps; aethalioid forms may approach the sporangiate type; the characteristics and disposition of limey secretions may be altered; spore maturation may be checked, resulting in spore-like bodies which are much larger than fully matured spores. Cold weather, and particularly frosts, may induce similar alterations. Such variations are in large part responsible for the extensive synonymy found in the group. Great caution is indicated in describing as new specimens that are the result of such environmental responses. They are not 'abnormal'; they are natural responses of the organisms involved to particular stimuli and must be so regarded. Giving them taxonomic

status as named varieties serves only to complicate the nomenclature and to extend the meaning of the category variety beyond its legitimate significance" (Martin & Alexopoulos 1969).

I am also reminded of an extensive fruiting of *Cribraria intricata* growing on well decayed decorticated wood of a coniferous log on a moist ground site in a heavily wooded forest — collection *HWK* #2898, Moro Bay State Park, Bradley County, July 8, 1989, State of Arkansas, USA. This fruiting covered several feet beginning at the underside of the log near ground level that was still moist and had various stages of sporangial development including mature long-stalked sporangia. This fruiting consisted of thousands of sporangia that covered the sides of the coniferous log from bottom to top where it had dried more quickly because of more exposure to the drying action of air currents. The stalked sporangia at the bottom had longer stalks and as the sporangia progressed toward the top of the log the stalks became shorter and shorter until some sporangia at the top were almost sessile. This fruiting was the same species and not intermixed with several species that sometimes happens with species of *Cribraria*. This is another example of stalk length variability when subjected to moisture gradients in nature.

Badhamia rugulosa is another example of a species new to science that was collected over a 13 - year period in four states (Florida, Kentucky, Missouri, and Ohio) and was based on 59 ample collections (Keller & Brooks 1975). Since that time additional collections have been made in Arkansas, Tennessee, and Texas. This species was found most frequently on the bark surface of living *Juniperus virginiana* (Eastern Red Cedar) trees and species of *Vitis* (grapevine). The following quotation documents an observational field experiment that records the variation in fruiting body morphology over time. This location is near Fairborn, Greene County, Ohio, USA.

"A single living *J. virginiana* tree, one among many large red cedar trees, located at John Bryan State Park in the lower picnic grounds was observed over a two-summer period. *Badhamia rugulosa* regularly formed extensive fruitings on this tree after rainy periods. When fructifications were observed immediately following fruiting, they appeared bright orange with the peridium rugulose above. After exposure to rains, the calcareous peridium and capillitial system showed varying degrees of bleaching from dull orange to white, eventually the only trace of the former bright orange color being the brownish streaks at the base of the fructifications. The calcareous peridium had often assumed a smoother quite different appearance when compared to unweathered specimens" (Keller & Braun 1999).

There are many other examples that demonstrate this plasticity and variability of myxomycete fruiting bodies, but these will suffice for this discussion.

Moist chamber cultures are often used to survey a given area or to find additional myxomycete specimens from habitats that may be too dry, too wet, or temperatures too cold or have less than ideal environmental conditions to produce fruiting bodies. This culture technique has produced many new distribution records for different countries or specific areas and at the same time has yielded species new to science. However, caution is required because moist chambers often produce aberrant or fewer fruiting bodies and therefore may pose additional problems when single or only a few collections serve as this only source of a species new to science. Too often moist chamber cultures of bark from living trees result in specimens with different collection numbers based on the date harvested from the cultures. However, all of the collections have come from a single tree and a single location and fail to meet the criterion of multiple collections from multiple different locations.

6 The importance of type collections

The literature is full of too many examples of species new to science that lack sufficient quantity of collections to (1) prepare a good species description; (2) a type specimen with adequate fruiting bodies to study with light or scanning electron microscopy; (3) serve to compare closely related species to determine if morphological characters are distinct or intergrade. Therefore, multiple collections from multiple locations will not only provide better source material for selection of a holotype collection but also enough material for isotypes and paratypes. This takes more time and will avoid the "rush to publish" especially if additional collections are obtained on loan from herbaria to compare morphological characters.

The importance of selecting and the study of type specimens cannot be overemphasized. These are the nomenclatural and morphological standards used for the taxonomy of both fossil and living organisms. Paper descriptions of supposedly species new to science are not always reliable as a substitute for holotype collections. One of the best examples is *Badhamia ovispora* described by Marian Raciborski (1863 – 1917) who was Director of the Department of Botany and the Botanic Garden of the Jagiellonian University in Cracow, Poland (1912 – 1917). Raciborski had followed in the footsteps of the previous director Józef Thomasz Rostafiński (1850 – 1928), also a renowned myxomycologist. The holotype collection was deposited in KRAM Polish Academy of Sciences.

A specimen collected by Henry Aldrich (*HA* #13) August 21, 1964, near Nederland, Colorado, Boulder County, USA, on a decaying gymnosperm log was abundant and in excellent condition. The spores were unique in shape (reniform to allantoid, referred to as hotdog-shaped) and with ornamentation as raised plaque-like areas (see Keller et al. 1975, Figs. 5 – 8, 13). Specimens sent to other myxomycologists all resulted in a declaration that this was a species new to science. It took numerous requests and about three years to obtain Raciborski's holotype collection.

Instead of publishing a new species I waited to study the holotype and compare it to Raciborski's published description to determine accuracy: "Spores variable in shape, ellipsoidal, rarely spherical, (14.5 – 16.5) μm × (7.5 – 8.3) μm and smooth" (Keller et al. 1975). Scanning electron micrographs of the holotype spores (see Keller et al. 1975, Figs. 9 and 10) show the elongate hotdog-shape and raised plaque-like areas on the spore surface. These spore characteristics were not included in the Raciborski holotype description. The raised plaque-like spore ornamentation in *HA*#13 can clearly be seen under the oil immersion lens using light microscopy at approximately 1000 × and the hotdog shapes are obvious at lower magnifications (see Figs. 5 – 7, Keller et al. 1975). The freeze-etched preparation of spores from *HA* # 13 shows the raised plaque-like areas in detail (Keller et al. 1975, Fig. 13). The use of higher magnification optics, especially scanning and transmission electron microscopy at 3250 × to 9120 ×, provided ornamentation evidence not seen at lower magnifications. The majority of myxomycete species have spores that are spherical or nearly so, except for *B. ovispora* that has an elongate hotdog shape. This is a rare species seldom collected but recently found by Y. Mourgues at Le Monêtier-les-Bains, a commune in the Hautes-Alpes department in southeastern France, on bark of a decaying *Fraxinus* log at an elevation of 1500 m and illustrated by Poulain et al. (2011).

7 Monographic concerns and considerations

Monographic publications are becoming a thing of the past. My monograph of the genus *Perichaena* (Keller 1971) included 13 species, and two of these, *P. brevifila* (MycoBank #319341) and *P. reticulospora* (MycoBank #319342), represented species new to science. Please note that *Mycologia* now requires a registered accession number in MycoBank. There was then and even now a trend to publish species new to science as short papers, emphasizing numbers over a single publication with a section of a genus or the entire genus. It was unfortunate my doctoral dissertation was never published as a single publication but was split into several papers, for example, *P. brevifila* (Keller & Brooks 1971), *P. reticulospora* (Keller & Reynolds 1971), and the spore-to-spore cultivation of *P. depressa* and *P. quadrata* (Keller & Eliasson 1992), and several other papers that included different portions of the thesis (Schoknecht & Keller 1977; Keller & Everhart 2008).

The two species new to science were represented by only 12 collections (all in deep compacted leaf litter only in the fall months) from three states Georgia, Kansas, and Virginia in the case of *P. brevifila*, and only a single collection from the type locality for *P. reticulospora*. In the latter case, the spores were bordered reticulate and unique for the genus *Perichaena*, and that is true today, even though the number of new taxa have expanded the genus to 26 species (Lado et al. 2009). This means that the total number of species new to science in the genus *Perichaena* has doubled in the last 38 years and that represents quite a remarkable number. Most of the *Perichaena* species new to science more recently have come from previously unexplored semiarid and arid areas of the world (Lado et al. 2009).

The Keller unpublished thesis (1971) had scanning electron micrographs for most of the *Perichaena* species, and in the case of *P. reticulospora* (Keller & Reynolds 1971) these were the first SEMs published of a myxomycete species new to science. This was the beginning of a new era when scanning electron microscopy would begin to usher in the importance of fine structure and detail of spore ornamentation. It has been more than 43 years, and *P. reticulospora* is still only known from the type locality in a secondary forest at the University of San Carlos Biological Station, Philippines. I am certain this species will be found in other tropical countries of southeast Asia even though it will be difficult to collect in the field because of its small size and occurrence as scattered stalked sporangia on decaying leaf litter on the forest floor. The participants in ICSEM 8 should read the species description of *P. reticulospora* and look for it in tropical forest ecosystems on ground litter.

8 Importance of spore-to-spore agar culture and spore characters

Spore-to-spore agar cultures should be encouraged and attempted. There are many recent examples represented in the papers published by Diana Wrigley de Basanta and Carlos Lado from Madrid, Spain, that should be consulted and followed as a model for future culture work (Lado et al. 2007, *Didymium wildpretii*; Mosquera et al. 2003, *Licea succulenticola*; Wrigley de Basanta & Lado 2010, *Licea eremophila*; Wrigley de Basanta et al. 2009, *Didymium infundibuliforme*; Wrigley de Basanta et al. 2011, *Didymium operculatum*; Wrigley de Basanta et al. 2012, *Physarum atacamense*). The combination of light microscopy of life cycle stages grown in culture and excellent scanning electron micrographs of spores merit special consideration. All of these examples of species new to science were collected in the field then grown in agar culture from spore-to-spore. *Didymium operculatum* spores were described as globose, (9.5) 10 – 11 (– 12.5) μm in diameter, banded reticulate with 9 – 12 meshes per hemisphere and with a second reticulum underneath, visable through the meshes as seen by SEM. These spore characters alone are distinct and would separate this species from all other known described myxomycete species, but more importantly, the spore characters were stable, constant, and reliable as taxonomic key characters.

Spore ornamentation and spore size appears to be a stable character with less variation as noted in the above examples. However, are these spore characters or any character measurable by DNA techniques? Schnittler & Mitchell (2000) in Table 1 noted clustered-spored species (adhering together in groups of 2 to 40) versus free-spored (single spore) species. They questioned the clustered-spore character as possibly being the result of a single gene mutation. Their discussion noted 28 taxa that have clustered spores listed along with rarity status, number in spore clusters, and possible equivalent species. Nine *Badhamia* species with clustered spores do not have species equivalents in the table, and to the best of my knowledge, there has been no DNA study to help determine the degree of separation for any of these species. I have collected and studied some of these species (number of collections in parentheses): *Badhamia crassipella* (27), *Didymium synsporon* (22), *Minakatella longifila* (20), *Perichaena syncarpon* (25), and *Physarum synsporum* (14) (Keller & Braun 1999).

Badhamia crassipella is known from the states of California and Washington, U.S.A., and although listed in category 2 by Schnittler and Mitchell (2000) it actually falls in category 3 (more than 20 collections designated as more common taxa) with no known species equivalent. This species was described as new by Whitney and Keller (1982), and the plasmodiocarpous habit distinguishes this species from other members of *Badhamia*. This taxon's clustered spores (4 – 40) do not appear to be a single gene mutation.

Didymium synsporum is a corticolous myxomycete known from the states of Arkansas, Georgia, Kansas, Kentucky, Ohio, and Tennessee, USA, and from the bare bark surface of living *Juniperus virginiana* trees. Spores adhere in firm clusters of 4 – 25 (Keller & Braun 1999). This species is most closely related to *D. difforme* (listed as the equivalent species, (Schnittler & Mitchell 2000), but the capillitium of mostly branching and anastomosing threads differs from the mostly upright, less branching capillitial threads of *D. synsporon* attached above to the peridium and below to the base of the fruiting body.

Minakatella longifila is a corticolous myxomycete only known from the bark of living trees and grapevines located in the states of Illinois, Iowa, Kentucky, Missouri, Ohio, and West Virginia, USA, and in Italy and Japan. It is probably more common and widespread than once thought (Keller et al. 1973; Keller & Braun 1999). There is no equivalent species (Schnittler & Mitchell 2000), but it is listed as a clustered (4 – 14) spored species.

Perichaena syncarpon is found on leaf litter that has accumulated under shrubbery around buildings and among decaying leaves under herbaceous plants and grasses in iris or lily beds. It is known from the states of Iowa and Kansas, USA. The type locality, in Junction City, Kansas, has been destroyed as part of a housing development (Keller 1971). Spores adhere in clusters of 4 – 16; *P. depressa* is listed as the free-spored species equivalent by Schnittler and Mitchell (2000). The pseudoaethalioid to aethalioid fructifications with dehiscence by lobes along prominent ridges or irregular aerolae in *P. syncarpon* differs from *P. depressa* with crowded and depressed sporangia on decayed wood with circumscissile dehiscence at the margins. There is no species equivalent to *Perichaena syncarpon* in the genus *Perichaena*, suggesting that in this case it is not a single gene mutation.

Physarum syncarpon was described as new to science by Stephenson & Nannennga-Bremekamp (1990) based on two moist chamber developments of bark from a living *Juniperus virginiana* tree from the same locality of Nicholas County, West Virginia, USA. This is an example of a new species based only on moist chamber cultures from a single locality and possibly a single tree. The description is based on line drawings and light microscopy. Indeed, all of the five new species of Myxomycetes (*Arcyria bulbosa*, *Arcyria colloderma*, *Diacheopsis rigidifila*, *Diderma brunneobasalis*, and *Physarum synsporon*) appear to be from a single collection from a single locality (Stephenson & Nannennga-Bremekamp 1990).

It is interesting to note that I had made 13 field-collected specimens during the summers of 1974 to 1976 from six different localities in Adams, Clinton, Greene, and Montgomery counties, Ohio that I had set aside as a possible new species (Keller & Braun 1999). All of these collections had capillitial tubules filled with calcium carbonate, and this clearly suggested a species of the clustered-spored Badhamias. This was ample material to describe a species new to science, but a cursory study of Dr. Travis Brooks' collections of *Badhamia versicolor* revealed several specimens that appeared later to be named *P. synsporon*. Indeed, all of the clustered-spored *Badhamia* species are badly in need of monographic work that includes examination of type specimens and SEM of spore morphology. This represents a possible taxonomic research project to study the Brooks' Collection of Badhamias and the holotype collection of *B. versicolor*.

The category 1 rarity status (Table 1, Schnittler & Mitchell 2000) listed *Calomyxa synspora* (equivalent species: *C. metallica*) and *Trichia synspora* (equivalent species: *Trichia varia*) among others that do appear similar enough to suggest a single gene mutation. None of the 28 different clustered-spored taxa considered have been grown from spore-to-spore in agar culture.

9 Examples of myxomycete monographic papers

Examples of monographic papers published in the 1970s were noted by Keller (1996) that included species descriptions based on extensive field collections, examination of type specimens and representative specimens from different herbaria, and light microscopy for illustrations. Professor Dr. Donald T. Kowalski who collected in the Cascade mountain range of northern California to Washington, USA, concentrated on the nivicolous myxomycetes (snowline or montane myxomycetes). His monographs of *Lamproderma* (Kowalski 1970), *Lepidoderma* (Kowalski 1971), and *Diacheopsis* (Kowalski 1975) merit special consideration along with the *Echinostelium* monograph by Whitney (1980) that included SEMs. There has been a gap of nearly 40 years, but the best most recent example of a monographic treatment is "*A taxonomic evaluation of the stipitate Licea species*" by D. Wrigley de Basanta & C. Lado (2005). This study included the use of light microscopy, SEM, and the species evaluation and nomenclature based on 21 type specimens. The next generation of myxomycologists would do well to follow these examples of monographs and spore-to-spore cultivation on agar culture.

References

Braun, K. L. 1971. Spore germination time in *Fuligo septica*. The Ohio Journal of Science, 71: 304 – 309.

Clark, J. 2000. The species problem in the myxomycetes. Stapfia,73: 39 – 53.

Clark, J., Haskins, E. F. 2014. Sporophore morphology and development in the myxomycetes: a review. Mycosphere, 5 (1): 153 – 170.

Keller, H. W. 1971. The genus *Perichaena* (Myxomycetes): a taxonomic and cultural study. Ph. D. Dissertation. University of Iowa, Iowa City. 199.

Keller, H. W. 1996. Invited Inaugural Address, Biosystematics of Myxomycetes: A Futuristic View, Second International Congress on the Systematics and Ecology of Myxomycetes (ICSEM2), organized by Real Jardín Botánico, CSIC, Madrid, Spain, Abstract/Volume. 23 – 37.

Keller, H. W. 2012. Myxomycete history and taxonomy: highlights from the past, present, and future. Based in part on the Invited Key note Address given at the Seventh International Congress on the Systematics and Ecology of Myxomycetes (ICSEM 7), Recife, Brazil. Mycotaxon, 122: 369 – 387.

Keller, H. W., Aldrich, H. C., Brooks, T. E. 1973. Corticolous Myxomycetes II: Notes on *Minakatella longifila* with ultrastructural evidence for its transfer to the Trichiaceae. Mycologia, 62: 768 – 778.

Keller, H. W., Aldrich, H. C., Brooks, T. E., et al. 1975. The taxonomic status of *Badhamia ovispora*: a myxomycete with unique spores. Mycologia, 67: 1001 -1011.

Keller, H. W., Braun, K. L. 1999. Myxomycetes of Ohio: Their systematics, biology and use in teaching. Ohio Biological Survey Bulletin New Series, 13 (2): 1 -182.

Keller, H. W., Brooks, T. E. 1971. A new species of *Perichaena* on decaying leaves. Mycologia, 63: 657 - 663.

Keller, H. W., Brooks, T. E. 1975. Corticolous Myxomycetes III: a new species of *Badhamia*. Mycologia, 67: 1218 - 1222.

Keller, H. W., Eliasson, U. H. 1992. Taxonomic evaluation of *Perichaena depressa* and *Perichaena quadrata* (Myxomycetes) based on controlled cultivation, with additional observations on the genus. Mycological Research, 96: 1085 - 1097.

Keller, H. W., Everhart, S. E. 2008. Myxomycete species concepts, monotypic genera, the fossil record, and additional examples of good taxonomic practice. Special Volume based on papers presented at the Fifth International Congress on Systematics and Ecology of Myxomycetes. (ICSEM 5). Tlaxcala, Mexico, August 8 - 13, Universidad Autónoma de Tlaxcala. Revista Mexicana de Micología, 27: 9 - 19.

Keller, H. W., Everhart, S. E. 2010. Importance of Myxomycetes in biological research and teaching. Fungi, 3 (1): 29 - 43.

Keller, H. W., Reynolds, D. R. 1971. A new *Perichaena* with reticulate spores. Mycologia, 63: 405 - 410.

Keller, H. W., Schoknecht, J. D. 1989. *Fuligo megaspora*, a myxomycete with unique spore ornamentation. Mycologia, 81: 454 - 458.

Keller, H. W., Schoknecht, J. D. 1989a. Spore-to-spore culture of *Physarum spinisporum* and its transfer to *Badhamia*. Mycologia, 81: 631 - 636.

Keller, H. W., Schoknecht, J. D. 1989b. Spore-to-spore cultivation of a new wrinkled-reticulate-spored *Badhamia*. Mycologia, 81: 783 - 789.

Kowalski, D. T. 1970. The species of *Lamproderma*. Mycologia, 62: 411 - 620.

Kowalski, D. T. 1971. The genus *Lepidoderma*. Mycologia, 63: 490 - 516.

Kowalski, D. T. 1975. The genus *Diacheopsis*. Mycologia, 67: 616 - 628.

Lado, C., Mosquera, J., Beltra'n-Tejera, E., et al. 2007. Description and culture of a new succulenticolous *Didymium* (Myxomycetes). Mycologia, 99: 602 -

611.

Lado, C., Wrigley de Basanta, D., Estrada Torres, A., et al. 2009. Description of a new species of *Perichaena* (Myxomycetes) from arid areas of Argentina. Anales delJardín Botánico de Madrid, 66S1：63 – 70.

Macbride, T. H. 1922. The North American Slime-moulds. 2 ed. New York：Macmillan Co. 299.

Martin, G. W., Alexopoulos, C. J. 1969. The Myxomycetes. Iowa City：University of Iowa Press. 561.

Mosquera, J., Lado, C., Estrada-Torres, A., et al. 2003. Description and culture of a new myxomycete, *Licea succulenticola*. Anales del Jardín Botánico de Madrid, 60：3 – 10.

Poulain, M., Meyer, M., Bozonnet, J. 2011. Les Myxomycètes. Vol 2, éditépar Fédération Mycologique et Botanique Dauphiné-Savoie, Editor. Sévrier, France.

Schnittler, M., Mitchell, D. W. 2000. Species diversity in myxomycetes based on the morphological species concept-a critical examination. Stapfia, 71：55 – 62.

Schoknecht, J. D., Keller, H. W. 1977. Peridial composition of white fructifications in the Trichiales (*Perichaena* and *Dianema*). Canadian Journal of Botany, 55：1807 – 1819.

Stephenson, S. L. 2011. From morphological to molecular：studies of myxomycetes since the publication of the Martin and Alexopoulos (1969) monograph. Fungal Diversity, 50 (1)：21 – 34.

Stephenson, S. L., Nannennga-Bremekamp, N. E. 1990. Five new species of Myxomycetes from North America. Proceedings of the Koninklijke Nederlandse Akademie van Wetenschappen, 93 (2)：187 – 196.

Sturgis, W. C. 1913. The Myxomycetes of Colorado, part II. Colorado College Publications Science Series, 12：435 – 454.

Tamayama, M., Keller, H. W. 2013. Aquatic myxomycetes. Fungi, 6 (3)：18 – 24.

Whitney, K. D. 1980. The myxomycete genus *Echinostelium*. Mycologia, 72：950 – 987.

Whitney, K. D., Keller, H. W. 1982. A new species of *Badhamia*, with notes on *Physarum bogoriense*. Mycologia, 74：619 – 624.

Wrigley de Basanta, D., Lado, C. 2005. A taxonomic evaluation of the stipitate

Licea species. Fungal Diversity, 20: 261 – 314.

Wrigley de Basanta, D., Lado, C. 2010. *Licea eremophila*, a new myxomycete from arid areas of South America. Mycologia, 102: 1185 – 1192.

Wrigley de Basanta, D., Lado, C., Estrada-Torres, A. 2011. Spore to spore culture of *Didymium operculatum*, a new Myxomycete from the Atacama Desert of Chile. Mycologia, 103: 895 – 903.

Wrigley de Basanta, D., Lado, C., Estrada-Torres, A. 2012. Description and life cycle of a new *Physarum* (Myxomycetes) from the Atacama Desert in Chile. Mycologia, 104:1206 – 1212.

Wrigley de Basanta D., Lado, C., Estrada-Torres, A., et al. 2009. Description and life cycle of a new *Didymium* (Myxomycetes) from arid areas of Argentina and Chile. Mycologia, 10: 707 – 716.

Harold W. Keller

Professor, Botanical Research Institute of Texas, Texas 76107, USA.

Research area

Tree canopy biodiversity of primarily myxomycetes and fungi. Corticolous myxomycetes from Central and Southeastern USA.

Work experiences

2009 – present	Professor Emeritus, University of Central Missouri, Warrensburg;
2006 – 2008	Visiting Professor, University of Central Missouri, Warrensburg;
2008 – 2006	Adjunct Professor of Biology, Central Missouri State University, Warrensburg;
1990 – present	Research Associate, Botanical Research Institute of Texas, Fort Worth;
1990 – 1998	Adjunct Associate Professor, Department of Microbiology, University of North Texas Health Science Center, Fort Worth;
1983 – 1990	Associate Professor, Department of Biology, University of Texas at Arlington;
1982 – 1983	Associate Professor, Department of Biological Sciences, University of

	North Carolina at Wilmington;
1978–1982	Associate Professor, Department of Microbiology and Immunology, Wright State University School of Medicine, Dayton, Ohio;
1971–1978	Assistant Professor, Department of Biological Sciences, Wright State University;
1971	National Science Foundation Fellow, "Summer Institute in Systematics V, Origin and Measurement of Diversity" held at Smithsonian Institute, Washington, D.C.

Journal papers, books, book chapters, journal papers at symposia and congress proceedings altogether 108.

Myxomycetology: A Challenging and Inspirational Field

Yu Li

Engineering Research Center of Chinese Ministry of Education for Edible and Medicinal Fungi, Jilin Agricultural University, Changchun, Jilin Province, China

Abstract: Slime molds are eukaryotic, heterotrophic microorganisms. They are famous for their interesting life cycle. Slime molds have proved to be model experimental organisms for studies of evolution, mitotic cycle, the structural physiology and movement of protoplasm, morphogenesis, aging, reproduction, and a variety of other questions that challenge scientists. Single-celled slime mold was proved that they can construct networks of nutrient channeling tubes strikingly similar to the layout of the Japanese rail system. In addition, slime mold was reported to have enviable capability of coping with an unknown dynamic environment and was being developed to control autonomous robots. Slime molds are also an important model organism in the genome research. There are four species that have been completely sequenced. Slime molds are also important node in the food web, especially as food for insects. About 100 natural compounds including fatty acid, amino acid, alkaloids, naphthoquinone, aromatic compounds, terpenoid, esters and their derivatives were found in slime molds. On the other hand, slime mold can also cause trouble in agricultural production. They are found on strawberry and grasses, and influence the growth of these plants. Slime molds occur upon the substrates, frames and grounds in mushroom houses, and will decrease the quality and yield of mushrooms. All those considered above have caused so much focuses on slime molds and intrigued so many biologists over the years. It is definitely indicated Myxomycetology is a challenging and inspirational discipline the same as biology.

Key words: Slime molds; biology-inspired; robot control; genome; compound

1 Introduction

1.1 What is slime mold?

Slime mold or slime mould is a broad term describing some organisms that use spores to reproduce. Slime molds were formerly classified as fungi but are no longer considered part of that kingdom. Slime molds are eukaryotic, heterotrophic microorganisms having a noncellular and multinucleate creeping vegetative phase and a propagative spore-producing stage (Li 2007). They exist in cool, shady, and moist places in the woods where they are found on decaying logs, dead leaves, and other organic materials. Unlike bacteria and other microorganisms, they ingest their food by phagocytosis in a manner similar to the amoebas. The vegetative cells are unique in that they lack cell walls. However, when fruiting bodies are formed, cell walls are present in these structures.

1.2 The life cycle of slime molds

Slime molds are famous for their interesting life cycle which includes a multinucleate somatic phase known as a plasmodium and a reproductive phase that culminates in the reproduction of stationary sporophores containing walled spores, which makes them very different from all other organisms. They begin life as amoeba-like cells. These unicellular amoebae are commonly haploid and multiply if they encounter their favorite food, bacteria. These amoebae can mate if they encounter the correct mating type and form zygotes which then grow into plasmodia. Plasmodia contain many nuclei without cell membranes and can grow to be meters in size. The amoebae and the plasmodia engulf microorganisms. The plasmodium grows into an interconnected network of protoplasmic strands. The cytoplasm can stream, stop, and then reverse direction within each protoplasmic strand. When the food supply wanes or environmental conditions change, the plasmodium will migrate to the surface of its substrate and transform into rigid fruiting bodies or a sclerotium. Exactly why sporangia form instead of sclerotia is still not clearly understood. These fruiting bodies or sporangia will then release spores which hatch into amoebae to begin the life cycle again (Martin and Alexopoulos 1969; Fiore-Donno et al. 2010).

1.3 Taxonomy of slime molds

Slime molds are a polyphyletic group. They are originally placed in the Fungi kingdom. But today, they are classified in the kingdom of Protozoa. Slime molds comprise the following three groups, Myxogastria or myxomycetes: syncytial or plasmodial slime molds, Dictyosteliida or dictyostelids: cellular slime molds, and Protosteloids (Kirk et al. 2008).

There are still conflicts to be resolved in the classification of these groups. Recent molecular evidence shows that the first two groups are likely to be monophyletic and the protosteloids however to be polyphyletic. For this reason, scientists are currently trying to understand the relationships among these three groups (Lahr et al. 2012).

2 Progress of Myxomycetology

Myxomycetology has greatly developed in the past 100 years on taxonomy, cell biology, ecology, molecular phylogeny, and other interesting fields. Here I will just mention a few frontier and special fields on slime molds. They are genetics, molecular biology, genomics, and bio-mathematics, bio-control robots, social dilemma, plant pathogen, etc.

2.1 Genetics of slime molds

The first systematic study of plasmodium fusion was made on *Physarum polycephalum* by Gray (1945) who presented evidence that plasmodia from a given geographical region would fuse with one another but those from different regions would not. Subsequently, many researches focused on the genetic control of plasmodium fusion. In 1972, Collins and Haskins reported that plasmodial (somatic) fusion in a strain of *Ph. polycephalum* is controlled by four loci, each of which displays simple dominance. Two diploid plasmodia fuse with each other only if they are phenotypically or genotypically identical for all four fusion loci. Recently, Barrantes and his group used a next-generation sequencing approach to study the transcriptomic changes during the differentiation of *Physarum* at the single-cell level in 2012. The observed regulation patterns correlate well with previous studies on differential gene expression during the commitment to sporulation in the slime mold, particularly with respect to proteins involved in signaling and actin binding.

2.2 Genomics of slime mold

The first genome of slime mold was completely sequenced from *Dictyostelium discoideum* in 2005 (Eichinger et al. 2005). It comprises six chromosomes, an extrachromosomal element encoding rRNA genes, and a mitochondrial genome. The phylogenetic tree constructed based on whole genomes confirms that *Dictyostelium* diverged along the branch leading to the Metazoa soon after the plant-animal split. Despite the earlier divergence of *Dictyostelium*, many of its proteins are more similar to human orthologues than are those of *Saccharomyces cerevisiae*, probably due to higher rates of evolutionary change along the fungal lineage. Later, Heidel and his group sequenced another three genome sequences from *Dictyostelium discoideum* (DD), *D. fasciculatum* (DF) and *Polysphondylium pallidum* (PP). The genome size of these three different species are 35 Mbp, 33 Mbp, and 33 Mbp, respectively (Heidel et al. 2011).

2.3 Biology-inspired mathematics

Computer science and biology have been partners for decades. Biologists rely on computational methods to analyze and integrate large data sets, while several computational methods were inspired by the high level design principles of biological systems. Recently, these two directions have been converging. Thinking computationally about biological processes may lead to more accurate models, which in turn can be used to improve the design of algorithms. Traditionally, biologists leveraged computing power to analyze and process data (e.g., hierarchically clustering gene expression microarrays to predict protein function), and computer scientists used high-level design principles of biological systems to motivate new computational algorithms (e.g., neural networks). Rarely these two directions were coupled and mutually beneficial. In 2010, Tero proved that single-celled slime mold can construct networks of nutrient channeling tubes strikingly similar to the layout of the Japanese rail system, which was borrowed by the researchers to create a biology-inspired mathematical description of the network formation and guide network construction.

2.4 Biological network

Many organisms grow and forage as remarkably large networks. Networks explore the environment to locate scarce and spatially disjunct resources (Boddy et al. 2009). The mechanisms used by an individual to integrate disparate sources of information, coordinate growth, and thrive across heterogeneous habitats remain unknown. The slime mold *Physarum polycephalum* has been repeatedly characterized as "intelligent" (Nakagaki and Guy 2008; Nakagaki et al. 2007, 2000). It can connect two food sources using the shortest path in amaze (Nakagaki et al. 2000), and networks connecting multiple food sources find a good compromise between efficiency, reliability, and cost, comparable to optimized, man-made, transport networks (Tero et al. 2010). Comparisons of theoretically generated contraction patterns with the patterns exhibited by individuals of *P. polycephalum* demonstrate that individuals maximize internal flows by adapting patterns of contraction to size, thus optimizing transport throughout an organism. This control of fluid flow may be the key to coordinating growth and behavior, including the dynamic changes in network architecture seen over time in an individual (Alim et al. 2013).

2.5 Bio-control autonomous robots

Slime mold was reported to have enviable capability of coping with an unknown dynamic environment and was being developed to control autonomous robots. Shimizu and Ishiguro (2009) presented *Slimebot* that enables an amoeboid locomotion. In that article, particularly aiming at achieving design scheme for local sensory feedback, they focused on the long-distance physical interaction stemming from the protoplasm and deformable outer skin. Umedachi et al. (2010) present a soft-bodied amoeboid robot inspired by the true slime mold.

2.6 Social dilemmas

Most interestingly, slime molds are one of the few non-human farmers. They can carry bacteria to seed out new food populations, and also carry other non-food bacteria which inhibit the growth of non-farmer *D. discoideum* clones that could exploit the farmers' crops. This is the first evidence of some form of cultivation beyond dispersing and seeding their food crops (Brock et al. 2013).

2.7 Influence on agriculture

On the other hand, slime mold can also cause trouble in agricultural production. They are found on strawberry, grasses, which influence the growth of these plants. Slime molds occur upon the substrates, frames and grounds in mushroom houses, which will decrease the quality and yield of mushrooms.

3 Prospects

Slime molds have proved to be valuable experimental organisms, not only for mycologists, but also for geneticists, molecular biologists, cytologists, biochemists, biophysicists, and bioinformatists. They are model experimental organisms for studies of evolution, mitotic cycle, the structural physiology and movement of protoplasm, morphogenesis, aging, reproduction, and a variety of other questions that challenge scientists. Many useful compounds are found in slime molds, including fatty acid, amino acid and some others. Slime molds may be new resources for natural medicinal products. Maybe we should do something to use slime mold for biomedicine production. Most important, myxomycetology not only draws mycologists' attention, but also arouses interests of researchers outside of biology. The future of Myxomycetology is bright to generate new ideas, new methods, new products, etc. Join us, and let your dreams come true.

All things considered have caused so much focuses on slime molds and intrigued so many biologists over the years. It is definitely necessary to diverge Myxomycetology as a discipline from biology, which will promote the development of biological science.

References

Alim, K, Amselema, G., Peaudecerf, F., et al. 2013. Random network peristalsis in *Physarum polycephalum* organizes fluid flows across an individual. PNAS, 110(33): 13306 – 13311.

Barrantes, I., Leipzig, J., and Marwan, W. 2012. A next-generation sequencing approach to study the transcriptomic changes during the differentiation of

Physarum at the single-cell level. Gene Regulation and Systems Biology, 6: 127 – 137.

Boddy, L., Hynes, J., Bebber, D. P., et al. 2009. Saprotrophic cord systems: Dispersal mechanisms in space and time. Mycoscience, 50(1):9 – 19.

Brock, D. A, Read, S., Bozhchenko, A., et al. 2013. Social amoeba farmers carry defensive symbionts to protect and privatize their crops. Nature Communication, 4: 1 – 7.

Collins, O. R., Haskins, E. F. 1972. Genetics of somatic fusion in *Physarum polycephalum*: the PpII strain. Genetics, 71: 63 – 71.

Eichinger, L., Pachebat, J. A., Glöckner, G., et al. 2005. The genome of the social amoeba *Dictyostelium discoideum*. Nature, 435: 43 – 57.

Fiore-Donno, A. M., Nikolaev, S. I., Nelson, M., et al. 2010. Deep phylogeny and evolution of slime moulds (Mycetozoa). Protist, 161(1): 55 – 70.

Gray, W D. 1945. The existence of physiological strains in *Physarum polycephalum*. American Journal of Botany, 32: 157 – 160.

Heidel, A. J., Lawal, H. M., Felder, M., et al. 2011. Phylogeny-wide analysis of social amoeba genomes highlights ancient origins for complex intercellular communication. Genome Research, 21(11): 1882 – 1891.

Kirk, P., M., Cannon, P. F., Minter, D. W., et al. 2008. Dictionary of the Fungi. 10th ed. Wallingford: CABI. 765.

Lahr, D. J. G., Grant, J., Nguyen, T., et al. 2012. Comprehensive phylogenetic reconstruction of Amoebozoa based on concatenated analyses of SSU – rDNA and Actin Genes. PLoS One, 6: e22780.

Li, Y. 2007. Flora of Fungi in China (Myxomycetes I). Beijing: Science Press.

Martin, G. W., Alexopoulos, C. J. 1969. The Myxomycetes. Iowa City: University of Iowa.

Nakagaki, T., Makoto, I., Ueda, T., et al. 2007. Minimum-risk path finding by an adaptive amoebal network. Physics Review Letters, 99(6):068104.

Nakagaki, T., Guy, R. 2008. Intelligent behaviors of amoeboid movement based on complex dynamics of soft matter. Soft Matter, 4(1):57 – 67.

Nakagaki, T., Yamada, H., Tóth, A. 2000. Maze-solving by an amoeboid organism. Nature, 407(6803):470.

Shimizu, M., Ishiguro, A. 2009. An amoeboid modular robot that exhibits real-time adaptive reconfiguration. In: Proceedings of the 2009 international conference on intelligent robots and systems (IROS'09), St. Louis, USA. 1496 – 1501.

Tero, A., Takagi, S., Saigusa, T., et al. 2010. Rules for biologically inspired adaptive network design. Science, 327(5964): 439 – 442.

Umedachi, T., Takeda, K., Nakagaki, T., et al. 2010. Fully decentralized control of a soft-bodied robot inspired by true slime mold. Biological Cybernetics, 102: 261 – 269.

Yu Li

Yu Li is a professor at Jilin Agricultural University, an academician of the Chinese Academy of Engineering, a Foreign Academician of the Russian Academy of Agriculture, a doctoral supervisor and National Excellent Educator and Science and Technology Worker.

Yu Li received his master degree of Science from the Chinese Academy of Science, and a Ph. D. from Tsukuba University of Japan. He is the Chairman of the first session of the International Institute of Medicinal Fungi, the chief editor of *International Medicinal Mushroom*, a National Young and Middle-aged Expert with great contribution. He was awarded the Special Government Allowance of the State Council. He is the first Chinese who named new species to slime molds nomenclature, also the first one who exploited the systematic taxonomy of slime molds at different taxa of genera, families, even orders. He has trained a great group of mycologists.

Resource Allocation and Morphogenesis during Fructification in Myxomycetes

Indira Kalyanasundaram

University of Madras, Karnataka, India

Abstract: Sporangial development in the Myxomycetes follows a standard pattern. After the sporangial initial has attained its full size, continuous loss of water leads to the formation of vacuole-like spaces within the protoplasmic mass.

In the Physaraceae, mostly excretion of unwanted material into these spaces gives rise to the capillitium. The stipe is formed by constriction of the outer membrane as the protoplasm moves up, and this space gets filled with refuse matter.

In the Didymiaceae and in the Stemonitales, material is secreted into the spaces to form the capillitium, the process being more refined in the Stemonitales. Thus a part of the resources would go into the making of the capillitium.

It is in the Trichiales that a considerable proportion of the resources is diverted towards the formation of the extensive, beautifully ornamented, hygroscopic capillitium and sometimes even the stipe, and the synthesis of various pigments. Considering the variety of these accessory structures, it seems worthwhile focusing attention on sporangial morphogenesis in this group.

Key words: Myxomycetes; resource allocation; capillitium; Trichiales

1 Introduction

With regard to Myxomycetes, much wonder has been expressed, by biologists laymen alike, at the sudden transformation of a motile, amoeboid slimy plasmodium into miniature fructifications of exquisite sculpture, with spores strongly resembling those of fungi. However, few people have cared to investigate the details of the process of this transformations. I may say that there has not been a detailed study of the life cycle of a single species to rival that of Wilson and Cadman, after the year of 1928. One reason might have been the tedious procedures involved in making the preparation for microscopical study. However, even with rapid improvements in technology, not much attention has been paid to this area, as is evident from a recent review in which emphasis is on Taxonomy rather than development processes (Clark & Haskins, 2014).

Generally speaking, the formation of sporangiate fructifications in Myxomycetes follows a standard pattern. When the sporangial initial becomes visible on the substratum, it has already attained its full size and is enclosed in a firm membrane or ectoplasmic gel. Further development entails a continuous but gradual loss of water from the protoplasm. This does not cause shrinkage as the membrane maintains the shape and size, but leads to the formation of vacuole-like spaces within the sporogenous protoplasm. The orientation of these spaces may differ in the different taxa and seems at times to be genetically controlled.

Even as these spaces grow and coalesce to form a network, material is secreted into these spaces to form the capillitial threads. These materials may fill up the cavities and fuse with membrane lining the cavity, giving rise to hollow or solid threads, depending on the amount of material secreted and their relationship to the membrane during development (Fig. 1).

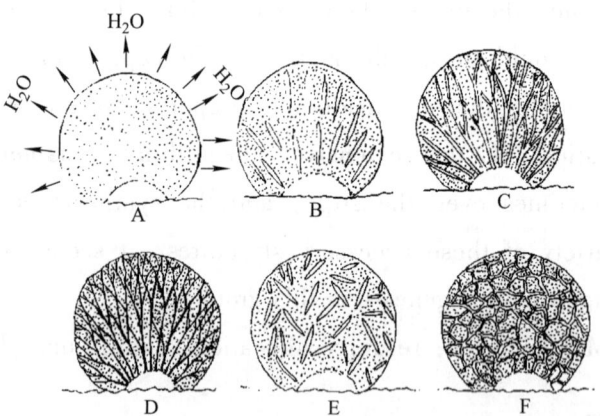

Fig. 1 A – D, Diagrammatic representation of Capillitial development, where capillitium extends from base to periphery; E and F, Irregularly reticulate development.

While this process of capillitium formation is going on, the sporogenous protoplasm goes through the processes of nuclear division and cleavage to form the spores (Fig. 2). How much of the protoplasm can be utilized for spore formation, would depend on the demands of the accessory structures in the major orders.

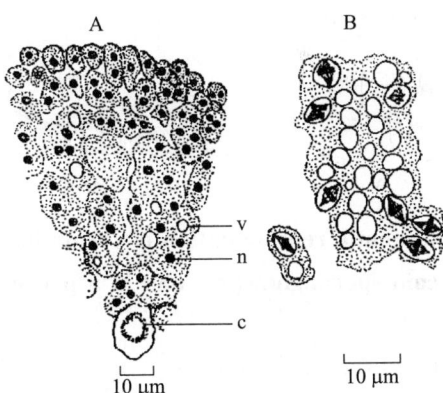

Fig. 2   Cleavage and nuclear division (From Indira 1969).

A, Cleavage in sporangium of *S. herbatica*; sector of a cross-section showing vacuoles (v), nuclei (n) and columella (c); B, Nuclear division, showing intranuclear spindles at metaphase.

2   Physarales

In the Physarales, especially in the family Physaraceae, the process is rather crude. It is usually unwanted material that goes into the formation of the accessory structures, so we may call it a process of excretion rather than secretion. The excess of calcium goes into the columella, the capillitium, onto the peridial surface and even into the stipe in species like *Physarum. melleum*, *P. citrinum* and *P. tenerun*.

In species with non-calcareous stipes, all kinds of refuse matter including fungal spores go to fill up the stipe, the outer membrane of which is continuous with the peridium. When sporangia develop in culture on agar media in such species, there being no ingested refuse matter to fill up the stipe, the stipe often remains weak and membranous.

It is not known in what form the calcium is excreted in these taxa, but its transformation into calcium carbonate seems to depend on contact with atmospheric carbon dioxide. In cultures where there is excessive moisture during fructification, the sporangia fail to turn white as they do when there is normal drying.

The orientation of the spaces that lead to the capillitial threads in the physaraceous species would depend on the shape of the sporangial heads. When these are globose or subglobose, they would radiate from the base. If flattened and disc-like, or tubular and convoluted, or in a plsamodiocarp, they would extend across, connecting the dorsal and ventral surfaces (Fig. 3).

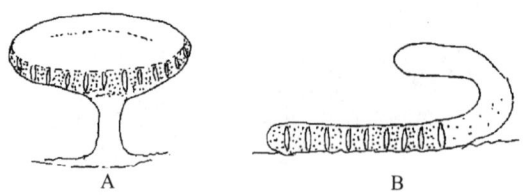

Fig. 3　Diagrammatic representation of formation of capillitial spaces in a discoid sporangium(A) and in a plasmodiocarp(B).

In the Didymiaceae the process is a little more refined. The capillitial threads generally radiate from the base, which often forms a columellar swelling, towards the peridium. They are often dichotomously branched, slender, apparently solid, lightly to strongly melanised in the different taxa (Fig. 4). Peridial lime is beautifully crystalline, but the stipe contains refuse matter as in the Physaraceae.

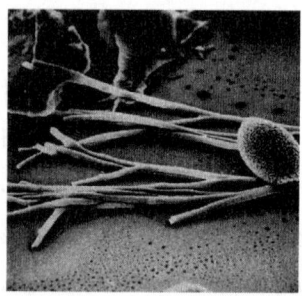

Fig. 4　SEM photograph of capillitialy threads of *Didymium*.
SEM, scanning electron microscopic.

Here again, most of the protoplasmic material goes into the making of spores, but one finds occasional exceptions. I may cite the case of *Didymium flexuosum* which has laterally compressed plasmodiocarps. Here the capillitium has many large, spore-like thickenings which could only have formed from sporogenous protoplasm (Kalyanasundaram 1978; Fig. 5).

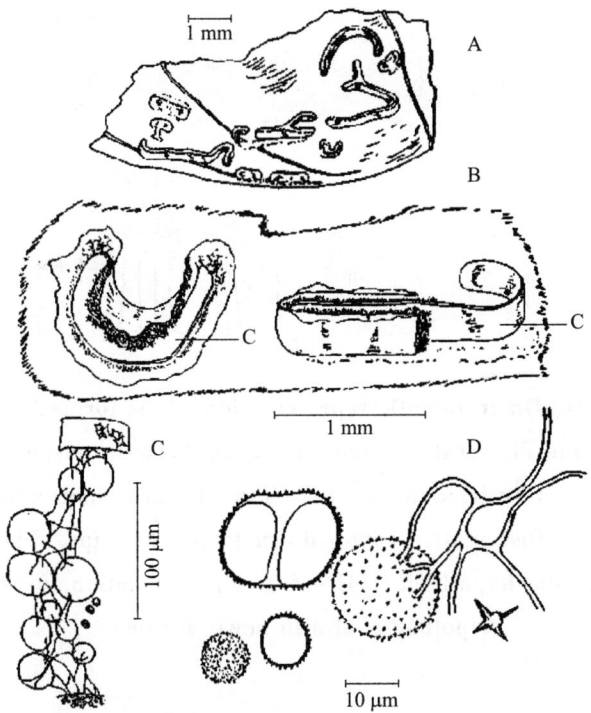

Fig. 5 Drawing of *Didymium flexuosum*, showing spore-like thickenings on capillitium (From Kalyanasundram 1975).

A, Plasmodiocarps on the edge of a leaf; B, Dehisced plasmodiocarps showing ribbon-like columella; C, Capillitium with vesicles, and a few spores; D, Capillitial thread, vesicles, lime crystal and spores.

3 Stemonitales

The process of differentiation is more refined in the Stemonitales. There seems to be no extrusion of unwanted material here as in the Physarales. The stipe is very different from that of the Physarales, in that it is formed internally and is homologous with the capillitium. Even the hypothallus is secreted from the sporogenous protoplasm and is continuous with the stipe (Fig. 6).

I can speak with some authority on this, having made a detailed study of the life cycle of *Stemonitis herbatica* nearly 50 years ago (Indira 1969; Fig. 7). The delicate, feather-like capillitium of the Stemonitales, culminating in the beautiful "surface-net" of the genus *Stemonitis*, would take up a fair proportion of the resources. In the simpler Lamprodermas, the capillitial threads arise from the small columella and are dichotomously branched towards the peridium. I think dichotomous branching is the simplest and most direct way of formation of vacuolar spaces that

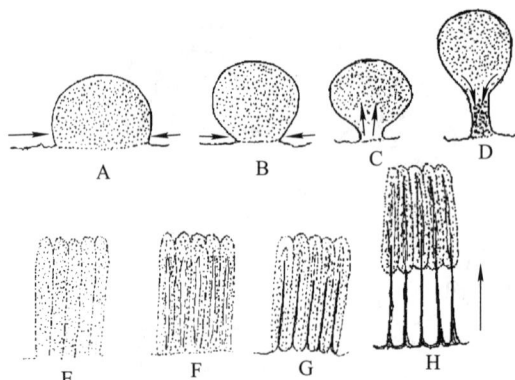

Fig. 6 A – D, Diagrammatic representation of the formation of stipitate sporangium in the Physarales. Constriction at the base pushes sporangium in the Physarales. Constriction at the base pushes sporangial protoplasm upwards, while refuse matter comes down to fill the stipe. The outer wall is continuous with the hypothallus. E – H, Stipe formation in *Stemonitis*. The hypothallus also in newly formed.

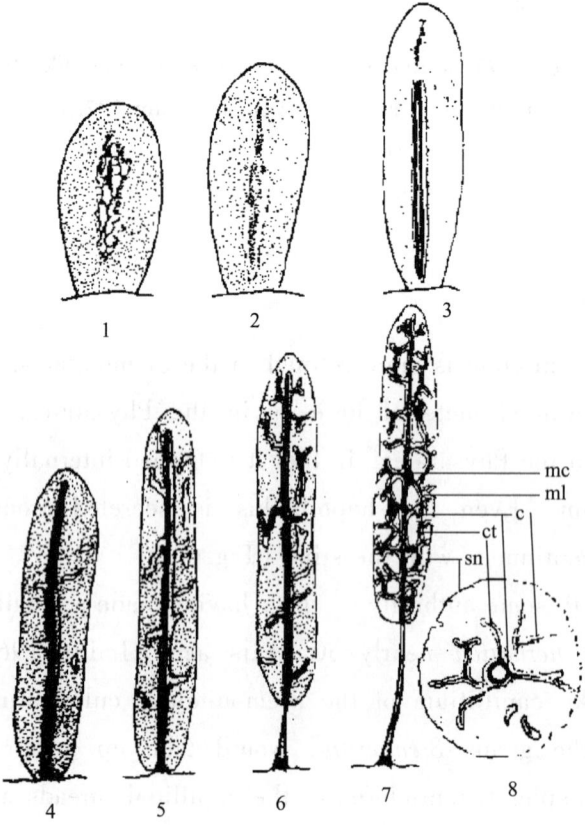

Fig. 7 Diagrammatic representation of capillitial formation in *Stemonitis*

(From Kalyanasundaram 1977).

radiate from the base of a sphere towards the periphery. It can be seen in many taxa with globose, subglobose or ovoid sporocarps, as already mentioned with regard to the Didymiaceae. One sees this in the simplest form in the Clastodermas. In the Stemonitales it is very common in *Lamproderma*. In *Comatricha* the branching is more complicated, and often one sees the beginnings of a surface net, that is so characteristic of *Stemonitis*.

The stipe and capillitium seem to be made of the same galactosamine polymer that strengthens the spore walls, and are heavily melanised (Paramasivan 1990). The melanin in this species has been found to be Dopa Melanin in the synthesis of which the reddish pigment L – DOPA (L – 3 – (3,4 – dihydroxyphenyl)alanine) is an intermediate (Loganathan and Kalyanasundaram 1999). This may be true of other myxomycetes as well, since many of them show a reddish tinge during development.

Unlike in the Physarales, the membrane covering the sporangial initial is not tough but very thin, giving rise to the evanescent peridium.

Spore ornamentation in the reticulately spored species is intriguing, the ridges being borne on rows of pillars, and the process of its formation is worthwhile investigating (Fig. 8).

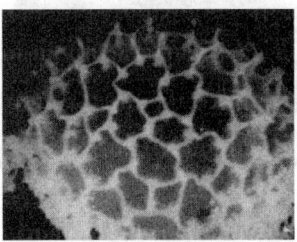

Fig. 8 SEM photograph of spore of *Stemonitis*.

My study of sporangial development of *S. herbatica* was based on a combination of squash and paraffin section preparations for light microscopical study. A paper presented at ICSEM 7 in Brazil, based on an electron microscopical study of development in *Lamproderma columbinum*, only seems to confirm my findings (Yajima 2011).

4 Liceales

In some of the Liceales, considerable resources seem to go into the making of the peridium, which is tough in the simple taxa, and beautifully sculpted in the basketlike Cribrarias with their peculiar dictydine granules.

5 Trichiales

It is in the Trichiales that sporangial morphology attains a peak, demanding considerable allocation of resources for the formation of accessory structures. The most marvellous is the hygroscopic capillitium, which expands gloriously in varied hues on dehiscence. It certainly calls for special synthesis of materials of which we do not even know the chemistry, despite several chemical's studies on Myxomycetes (Ishibashi 2005). Chemistry apart, the sculpting of the threads having a twist within a twist, spiral thickenings scattered with spines, longitudinal ridges supporting the spirals, often a hollow interior — all these are truly mind-boggling(Figs. 9—11). Hence a developmental study on this group would be truly rewarding.

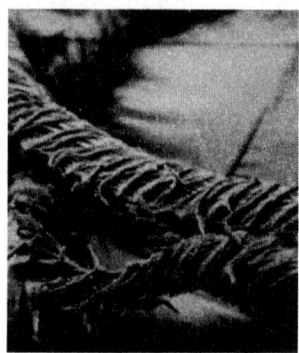

Fig. 9 SEM photograph of capillitial thread of *Trichia*.

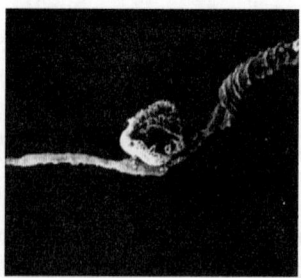

Fig. 10 SEM photograph of tip of capillitial thread of *Trichia*, showing inner core, and spore.

The formation of stipe in genera like *Arcyria* and *Hemitrichia* would call for a considerable sacrifice, because the 'cells' that fill up the stipe cannot be anything

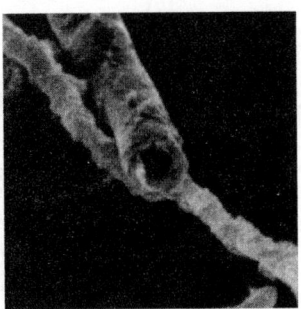

Fig. 11 SEM photograph of broken capillitial thread of *Perichaena* showing double wall and hollow interior.

other than aborted spore initials (Fig. 12).

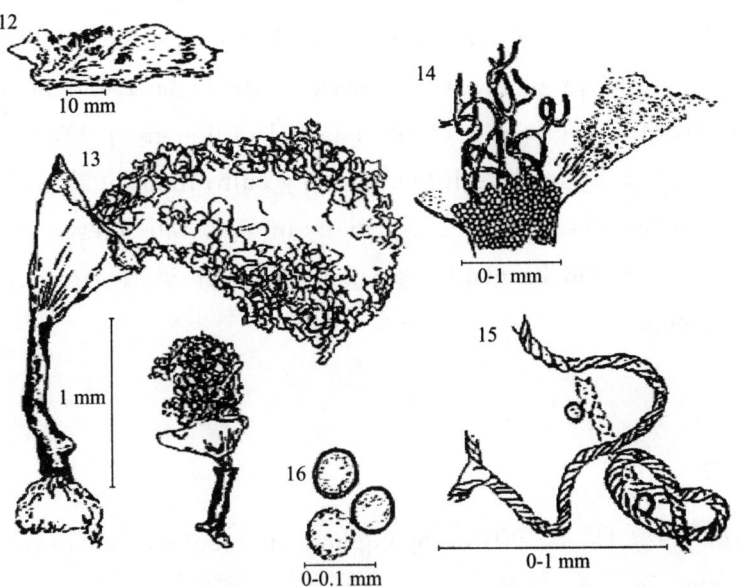

Fig. 12 Drawing showing stipe of *Hemitrichia*, filled with spore-like cells (From India 1968).

12, Sporangia on a piece of wood; 13, Two sporangia enlarged, to show variation in size; 14, Basal region of the sporangium, showing vesicles filling the stipe, and capillitial threads rising from the stipe; 15, Capillitial threads with spores; 16, Spores.

Equally intriguing are the spores with reticulate ornamentation, where one finds a reticulum within a reticulum. The material comprising these wing-like ridges seems to be different from that of the spore wall, and merits detailed study (Fig. 13).

The pigments, which colour spores and capillitium alike, also offer much scope for study. Many have been worked out by Steglich and co-workers in Germany (Gill

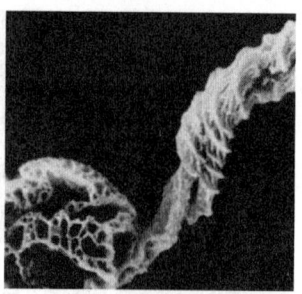

Fig. 13 SEM photograph of spore of *Trichia*.

& Steglich 1987), and by some Japanese workers, but there are many more. These pigments seem to be synthesized after melanin, which has been found in spores of all bright-spored species studied by us (Kalyanasundaram et al. 1994).

　　Thus the highest resource allocation for accessories of the fructification and the most elaborate morphology of these accessories, are to be found in the Trichiales. Unfortunately, morphogenesis is least investigated in this group. While it is true that this group does not easily lend itself to laboratory cultivation, it should be possible to find stages of development in nature, especially in temperate forests. And anyone who makes such a study would be making a significant contribution to our knowledge of this wondrous group.

References

Clark, J., Haskins, E. F. 2014. Sporophore morphology and development in the myxomycetes: a review. Mycophere, 5: 153 – 170.

Gill, M., Steglichm, W. 1987. Pigments of Fungi (Macromycetes). New York: Springer-Verlag. 122 – 123.

Indira, P. U. 1969. The life cycle of *Stemonitis herbatica*. 11. Transactions of the British Mycological Society, 56: 251 – 259.

Ishibashi, M. 2005. Isolation of bioactive natural products from Myxomycetes. Medical Chemistry, 1: 575 – 590.

India, P. U. 1968. Some slime moulds from Southern India-VIII. Journal of Indian Botanical Society, 47: 155 – 186.

Kalyanasundaram, I. 1975. Some slime moulds from Southern India-IX. Kavaka, 3: 41 – 54.

Kalyanasundaram, I. 1977. Capillitial development in *Stemonitis*. In: Subramaniam, C. V. (ed). Taxonomy of Fungi. 9 – 13.

Kalyanasundaram, I. 1978. *Didymium flexuosum*: an SEM study. Mycotaxon, 7: 125 – 129.

Kalyanasundaram, I., Menon, L., Loganathan, P. 1994. Occurrence of melanin in bright-spored Myxomycetes. Cryptogamie Mycology, 15: 229 – 237.

Loganathan, P., Kalyanasundaram, I. 1999. The melanin of the myxomycete *Stemonitis herbatica*. Acta Protozoologica, 38: 97 – 103.

Paramasivan, P. 1990. Some physiological aspects of the life cycle of *Stemonitis herbatica*. Ph. D. thesis, University of Madras, India.

Wilson, M., Cadman, E. J. 1928. The life-history and cytology of *Reticularia lycoperdon* Bull. Transactions of The Royal Society of Edinburgh-earth Sciences, 55: 555 – 603.

Yajima, Y. 1911. A study of sporophore development in *Lamproderma columbinum*. Abstracts of the VII International Congress on Systematics and Ecology of Myxomycetes. Recife, Brazil. 81.

Indira Kalyanasundaram

Professor, University of Madras, Karnataka, India.

Subjects of research

Myxomycetes; Foodgrain Storage Fungi&Mycotoxins; Medical Mycology; Ethnobotany.

Subjects taught

Mycology; Microbiology; Plant Ecology (post-graduate level),

Publications

37 papers about Myxomycetes, others 26 papers.

中国黏菌生物学研究进展

王琦，李姝

吉林农业大学食药用菌教育部工程研究中心，吉林长春

摘要：中国的黏菌研究工作开始于20世纪30年代，以真黏菌及细胞状黏菌为主要研究对象，在真黏菌的生物学研究方面，我国学者对多种黏菌的实验室培养方法进行了比较研究，并在此基础上完成了多种黏菌的生活史观察，从个体发育角度对黏菌的分类研究与系统关系进行了分析。

关键词：湿室培养；有饲培养；原生质团；个体发育

Biological Research of Myxomycetes in China

Qi Wang, Shu Li

Engineering Research Center of Chinese Ministry of Education for Edible and Medicinal Fungi, Jilin Agriculture University, Changchun, China

Abstract: The study of slime mold was begun from the 1930s including Myxomycetes and Dictyostelids and many books and papers about its taxonomy, phylogeny and chemistry have been published since 1990s in China by Chinese researchers. Based on the lab culture of myxomycetes, life-cycles of species were obtained, and the differences of ontogenetic development among species were discussed for taxonomy and phylogeny.

一、前言

黏菌（slime mold）包括真黏菌、网柄菌、集胞菌和根肿菌4个类群（Kirk et al. 2008），黏菌常见于森林阴凉湿润之地，可生长在土壤、腐木、落叶、活体动物、

植物及真菌上,分布于世界各地。黏菌具有包含大量孢子的繁殖体,形似真菌;在黏菌孢子萌发后形成的游动胞、黏变形体及营养生长阶段的原质团又体现了黏菌与动物相似的活动特性(李玉等 2008;Singh 1947;Cavender 1969a;阿力索保罗等 2002)。基于黏菌的生活学特征,黏菌的分类地位一直以来都存在着较大争议,先后被归属到菌物界与原生动物界。另外,在细胞遗传学、生理生化、活性成分开发等多方面都具有重要的理论研究和应用价值,常常被用作研究细胞衰老、癌症的模型生物(Clark 1984;Clark & Lott 1989)。可以说,黏菌的生物学研究为其分类归属、应用开发等方面工作奠定了基础。

二、黏菌的实验室培养

真黏菌营养生长阶段是独立生活、多核非细胞结构、仅有表面质膜而无细胞壁、能变形移动和摄食有机物的一团原生质,称为原生质团。从营养生长阶段转入繁殖阶段时,原生质团转变为一个或一群非细胞结构的孢子体,孢子体的表面或内部产生孢子。在其整个生活史和循环中以双倍体阶段为主。由于黏菌的子实体微小,在野外不易采集,因此,实验室培养成为获取黏菌的重要途径。

(一) 湿室培养

1933 年,Gilbert 和 Martin 采用湿室培养的方法在实验室条件下获得了大量黏菌(Gilbert & Martin 1933),不仅培养条件简便可控,不因季节变化而产生影响,并且可以补充野外记录,便于观察黏菌原质团到子实体的发育成熟过程,比较个体发育中在形态上的差异性,为黏菌的生物学研究提供了基础条件。

我国最早研究黏菌湿室培养的是周宗璜,在实验室条件下培养出了 18 种黏菌(周宗璜等 1981)。此后湿室培养逐步成为研究黏菌物种多样性及其生活习性的重要手段(赵日丰 1983;潘景芝 2009;刘福杰等 2010;徐美琴等 2006),研究者们通过此方法培养并报道 9 个中国新纪录种:*Diacheopsis metallica*, *Licea pedicellata*, *Physarum auriscalpium*, *Comatricha solitaria*, *Licea kleistotbolus*, *Licea synsporos*, *Perichaena vittate*, *Diderma donkii*, *Enerthenema intermedium*, *Paradiacheopsis rigida*, *Collaria arcyrionema*(袁海滨和陈双林 1996;陈双林和李玉 1995;李新宇 2002)。

(二) 有饲培养

黏菌的有饲培养也是获得大量研究材料的重要途径之一,1998 年,王琦等利用玉米琼脂培养基对团毛菌目黏菌的主要属种进行了有饲培养研究(王琦等 1998)。2003 年,史立平对绒泡菌目和团网菌目 8 种黏菌进行了实验室有饲培养与研究(史立平 2003),在培养基上观察到完整的孢子到孢子的生活循环,并且指出培养的子实体与野生的差别很小。2004 年,朱鹤等对团毛菌目黏菌个体发育的培养条件进行了研究(朱鹤和王琦 2004)。2006 年,刘朴等(刘朴和王琦 2006)研究了细弱绒泡菌的孢子到孢子的生活史,其中的孢子萌发细节通过显微观察技术得到,并且在燕麦培养基上,形成了成熟的子实体,与野生子实体比较差别不大,具有可育性。此方法为研究黏菌生活史中的个体发育特征提供了重要途径。

(三)液体发酵培养

国内有关黏菌的原质团的液体培养较少(刘士德和张建华 2004;刑苗和曾宪录 2000)。2010 年,刘福杰尝试对淡黄绒泡菌(*P. melleum*)和全白绒泡菌(*P. globuliferum*)的显型原质团进行了液体培养(刘福杰 2010),并以培养出颗粒状原质团生物量的增长和产生挂壁现象的程度为参考指标,对培养条件进行了初步研究。2011 年,谷硕等对绒泡菌目 8 种黏菌:淡黄绒泡菌(*P. melleum*)、全白绒泡菌(*P. globuliferum*)、扁绒泡菌(*P. compressum*)、灰绒泡菌(*P. cinereum*)、小绒泡菌(*P. pusillum*)、两瓣绒泡菌(*P. bivalve*)、灰堆钙丝菌(*Badhamia cinerascens*)、黑柄钙皮菌(*Didymium nigripes*)原质团进行了液体培养(谷硕 2011),以最适宜的燕麦液体培养基为基础,对培养基 pH、培养温度和转速进行优化研究。

三、黏菌的生活史循环

1858 年,de Bary 对一些黏菌的生活史进行了详细研究(de Bary 1858),通过大量实验来揭示黏菌的形态发生和生长发育各个阶段,并对黏菌子实体的两种发育方式和黏菌的基本繁殖方式做了详尽描述,这是有关黏菌个体发育研究最早的观察。1901 年,Lister 父女开始了对真黏菌生活史的研究(周宗璜 1981),我国学者对真黏菌生活史的研究则始于 21 世纪(史立平 2003;刘朴 2007;史立平和李玉 2003,2004,2005,2007,2008;史立平等 2006;Liu et al. 2008,2010;Chen et al. 2013),包括:针箍菌(*Physarella oblonga*)、圈绒泡菌(*P. gyrosum*)、黄柄钙皮菌(*D. iridis*)、细弱绒泡菌(*P. tenerum*)、扁绒泡菌(*P. compressum*)、煤绒菌(*Fuligo septica*)、多头绒泡菌(*P. polycephalum*)、灰团网菌(*Arcyria cinerea*)、全白绒泡菌(*P. globuliferum* Pers.)、小绒泡菌(*P. pusillum*)。

(一)孢子萌发

孢子萌发是黏菌开始生命活动的第一步,也是黏菌研究者关注的问题之一。

对于黏菌孢子萌发条件的研究,Gilbert 对黏菌孢子在水中的萌发率、萌发时间等问题进行了探讨(Gilbert 1929),同时,还研究了温度、pH 值和光照等环境因素对孢子萌发的影响。1937 年,Smart 发现孢子萌发受到营养条件、温度、氢离子浓度等因素影响(Smart 1937),其中,绒泡菌目的黏菌对营养条件具有较强的适应能力。1949 年,Elliott 指出树皮、木头、草叶的稀煎汁可刺激孢子萌发,且孢子萌发方式分为"孔出式"和"V 型开口式"(Elliott 1949)。2003 年,史立平等研究了不同稀煎汁和不同 pH 条件下 8 种黏菌孢子萌发情况,结果显示萌发率最高的是绒泡菌目黏菌孢子,在 pH 2.0~9.0 的条件下孢子均可萌发(史立平和李玉 2003)。

(二)游动胞与黏变形体

孢子萌发后可产生游动胞或黏变形体,在水分充足的环境下孢子释放出具有鞭毛的游动胞,而当水分减少时,游动胞则会变成黏变形体(李玉等 2007)。黏变形体形态结构与运动方式在种间有所差异,并且黏变形体的结构对变形运动方式具有一定影响(李晨等 2013)。

(三)原质团

原生质团是黏菌营养生长时期的主要存在形式,是由一层薄的质膜和胶黏质鞘包着的原生质,鞘上有微纤丝,它无结构,包含着小颗粒、液泡和其他物体,它没有形状大小的限制,为球形、扁平片状或扩展的纤细网状,常具鲜艳色彩,不断变形流动裹吞食物,最终形成子实体(李玉等 2007)。2005 年,王晓丽等对几种黏菌的显型原质团的一般形态、内含物、培养特性、减数分裂期进行了研究(王晓丽等 2005),结果表明:典型的显型原质团为扁平的网脉状结构,前端呈扇形,具有不同的颜色,为一团裸露的原生质,外无细胞壁,包有一层胶黏质的鞘,鞘含有微纤丝,并随着原质团在基物上爬行过程中脱落而留下痕迹,显型原质团内部是多核的,原质团内含有大量的石灰质,原质团易被其他微生物侵染,不同原质团中的内含物大致相同,针箍菌的减数分裂发生在孢囊形成时期、孢子形成前。2007 年,王晓丽等从煤绒菌显型原质团中提纯细胞核、核仁,诱导原质团形成菌核,并在透射电镜下观察,研究结果表明,细胞核具有中央核仁,核仁可以看到明显的纤维中心、致密纤维中心和颗粒结构,原质团中存在大量的黏液颗粒(王晓丽等 2007)。

黏菌营养物质有糖类和蛋白质类有机物(Chet et al. 1977; Kincaid & Mansour 1978; Knowles & Carlile 1978; McClory & Coote 1985)、燕麦(Nakagaki et al. 2000)、细菌(Konijn & Koevenig 1971)、真菌子实体(Emoto 1932; Madelin et al. 1975)等。原生质团可通过细胞质节律性流动、营养循环和化学信号对环境中的刺激做出应答(Dussutour et al. 2010),这种黏菌原生质团的趋向行为也引起了许多学者的注意并对此进行多种测试,如寻找食物间的最短距离(Nakagaki et al. 2000；Bonifaci et al. 2012)、预期周期事件的发生时间(Saigusa et al. 2008)及对食物营养的选择问题(Konijn & Koevenig 1971；Emoto 1932；Madelin et al. 1975；Dussutour et al. 2010；Latty & Beekman 2011；Lister 1888)。2010 年,谷硕对淡黄绒泡菌、全白绒泡菌、扁绒泡菌、小绒泡菌、两瓣绒泡菌、灰堆钙丝菌、黑柄钙皮菌的原生质团对细菌的吞噬行为进行了探究,发现黏菌原生质团对于细菌的吞噬形式在种间具有差异(谷硕 2011)。2013 年,姜宁对绒泡菌目 5 种黏菌进行糖类趋化性实验,结果表明,不同的黏菌对糖类具有不同的趋化性,其中煤绒菌趋向于葡萄糖和乳糖；淡黄绒泡菌趋向于半乳糖和麦芽糖；细钙丝菌、鳞钙皮菌和大孢钙皮菌趋向于葡萄糖和麦芽糖(姜宁 2013)。2013 年,宋晓霞对绒泡菌目的大孢钙皮菌、鳞钙皮菌、淡黄绒泡菌、细钙丝菌、煤绒菌和针籦菌 6 种显型原生质团在谷类、菇类和蔬菜类营养的选择性差异进行了比较,认为各原生质团本身的营养状况与其生境有关,隶属于钙皮菌科的大孢钙皮菌和鳞钙皮菌原生质团对食物代谢效率较高,偏好谷类食物中的燕麦；而隶属于绒泡菌科的淡黄绒泡菌、细钙丝菌、煤绒菌和针籦菌代谢效率较低,偏好菇类食物中的香菇(宋晓霞 2013)。

(四) 子实体发育

2003 年,史立平等在燕麦琼脂培养基上对针籦菌 $P.\ oblonga$ 的个体发育过程进行了研究(史立平和李玉 2003),并且培养出了具有可育性子实体的孢子。2011 年,谷硕等通过使用液体发酵和有饲培养相结合的方法,对绒泡菌属黏菌全白绒泡菌($P.\ globuliferum$)、扁绒泡菌($P.\ compressum$)、淡黄绒泡菌($P.\ melleum$)的原质团进行了孢子果的诱导,并在饥饿条件下,通过对光照和温度的调节,获得了绒泡菌属黏菌在不同培养条件下形成孢子果和菌核的最佳条件(谷硕等 2011)。

四、黏菌的个体发育研究

Kendrick 指出,个体发育是系统发育各进化阶段的重演,个体发育的不同类型也体现了系统发育的各个环节不同阶段和不同分支,使分类系统更好地体现了系统发育关系(Kendrick 1987)。1887 年,de Bary 根据真黏菌形成原质团,提出黏菌与原生动物关系更密切,强调了黏菌的动物属性,并称之为 Mycetozoa(菌虫)(de Bary 1887)。

1959年,Ross通过一系列个体发育研究的实验数据(Ross 1959),提出了发网菌目黏菌与腹黏菌的其他目在原质团类型及子实体发育方式上存在明显差异,发网菌原质团为隐型,子实体发育方式为基质层上型;腹黏菌原质团为显型或原始型,子实体发育方式为基质层下型,由此在1973年建立了发网菌亚纲,这一分类方案至今仍得到广泛认可。此后,黏菌的个体发育研究为其分类关系带来了陆续的调整。Alexopolous在完成刺轴菌目数种黏菌的个体发育研究后指出,黏菌的原质团特征与高级分类单元间具有特定联系,刺轴菌的原质团为原始型,子实体形成方式为基质层下型,与其他黏菌都不同(Alexopoulos 1960)。据此,Martin建立了刺轴菌目(Martin & Alexopoulos 1969)。同样,Farr也通过个体发育试验证明,白柄菌属(*Diachea*)的子实体发育方式为基质层下型,应归于绒泡菌目(Farr 1974),再一次表明,个体发育研究在黏菌系统分类研究中具有极其重要的作用。

2006年,王琦等利用有饲培养方法,对团毛菌目黏菌的盖碗菌属(*Perichaena*)、团网菌属(*Arcyria*)、半网菌属(*Hemitrichia*)、变毛菌属(*Metatrichia*)及团毛菌属(*Trichia*)7个种的个体发育进行了比较研究,发现孢丝或孢子表面纹饰不明显的种类较原始,表面纹饰明显的种类为发达类型(王琦等2006)。在此基础上,通过比较团毛菌目子实体形态结构,将团毛菌目黏菌子实体分为原始型、中间型及进化型,同时,将团毛菌目系统划分为两个发展方向,散丝菌科(Dianemaceae)和团毛菌科(Trichiaceae),且每一分支都有自原始形式向分化类型的发展过程(王琦和李玉2006)。

2013年,宋晓霞在实验室培养的团毛菌目、绒泡菌目和发网菌目14种黏菌的形态建成过程基础上,发现团毛菌目、绒泡菌目和发网菌目黏菌在原生质团的形成过程和生长方式上存在共性,发网菌目具有更高级可移动的珊瑚状结构,从四目对应的原生质团生活史策略及子实体形态结构的复杂程度来看,认为无丝菌目为较原始的类群,团毛菌目、发网菌目和绒泡菌目为较高级的有同源性的类群(宋晓霞2013)。

五、展望

黏菌的生物学研究为人们揭示了一种特殊类群的生命活动,研究者们更是受此启发将研究结果应用到其他领域,如何更透彻地解析黏菌的生物学行为及与此相关机制,将是今后黏菌生物学研究的重点问题。同时,以此为基础的个体发育研究也将为黏菌的分类系统提供新的理论支持。另外,在近70年的研究工作中,我国的黏菌研究得到了长足发展,涉及分类学、细胞学、分子生物学及化学等多个领域,为系统性地揭示黏菌的起源、进化奠定了坚实基础,但是,与国外相比，我国很多研究还处于起步阶段，在多年的研究基础上，我国学者对黏菌研究的见解，也需要今后的学者们继承与发扬。

六　致谢

本文由国家自然科学基金项目（31370065），国家科技支撑项目课题（2012BAC01B04）支持。

参考文献

Alexopoulos, C. J. 1960. Gross morphology of the plasmodium and its possible significance in the relationships among the myxomycetes. Mycologia, 52: 1 – 20.

Bonifaci, V., Mehlhorn, K., Varma, G. 2012. *Physarum* can compute shortest paths. arXiv:1106.0423v3, 2012.

Cavender, J. C. 1969a. The occurrence and distribution of Acrasieae in forest soils. I. Europe. American Journal of Botany, 56: 989 – 992.

Chen, X. S., Gu, S., Zhu, H., et al. 2013. Life cycle and morphology of *Physarum pusillum* (Myxomycetes) on agar culture. Mycoscience, 54: 95 – 99.

Chet, I., Naveh, A., Henis, Y. 1977. Chemotaxis of *Physarum polycephalum* towards carbohydrates, amino acids and nucleotides. Journal of General Microbiology, 102: 145 – 148.

Clark, J. 1984. Lifespans and senescence in six slime molds. Mycologia, 76: 366 – 69.

Clark, J., Lott, T. 1989. Age heterokaryon studies in *Didymium iridis*. Mycologia, 81: 636 – 638.

de Bary, A. 1858. Uber die Myxomyceten. Bot Zeit, 16: 357 – 358, 361 – 364, 365 – 379.

de Bary, A. 1887. Comparative Morpnology and Taxonomy of the Fungi, Mycetozoa, and Bacteria. Clarendon, Oxford.

Dussutour, A., Latty, T., Beekman, M., et al. 2010. Amoeboid organism solves complex nutritional challenges. Proceedings of National Academy of Science, USA, 107(107): 4607 – 4611.

Elliott, E. W. 1949. The swarm-cells of myxomycetes. Mycologia, 41: 141 – 170.

Emoto, Y. 1932. über die Chemotaxis der Myxomyceten-Plasmodien. Proceedings of the Imperial Academy of Japan, 8: 460 – 463.

Farr, M. L. 1974. Some new myxomycetes records for the neotropics and some taxonomic problems in the myxomycetes. Proceedings of Iowa Academy of Science, 81: 37 – 40.

Gilbert, F. A. 1929. Spore germination in the myxomycetes: a comparative study of spore germination by families. American Journal of Botany, 16: 421 – 432.

Gilbert, H. C., Martin G, W. 1933. Myxomycetes found on the bark of living trees. University of Iowa Studies (Natural History), 15:3 – 8.

Kendrick, B. 1987. 现代真菌学. 张素轩, 等, 译. 南京:南京林业大学出版社: 5 – 7.

Kincaid, R. L., Mansour, E. 1978. Chemotaxis toward carbohydrates and amino acids in *Physarum polycephalum*. Experimental Cell Research, 116: 377 – 385.

Kirk, P. M., Cannon, P. F., Minter, D. W., et al. 2008. Ainsworth & Bisby's Dictionary of the Fungi. 10th ed. CABI Bioscience, CAB International, UK.

Knowles, D. J. C., Carlile, M. J. 1978. The chemotactic response of plasmodia of the myxomycete *Physarum polycephalum* to sugars and related compounds. Journal of General Microbiology, 108: 17 – 25.

Konijn, T. M., Koevenig, J. L. 1971. Chemotaxis in Myxomycetes or true slime molds. Mycologia, 63: 901 – 906.

Latty, T., Beekman, M. 2011. Irrational decision-making in an amoeboid organism: transitivity and context-dependent preferences. Proceedings of the Royal Society, B, 278: 307 – 312.

Lister, A. 1888. Notes on the plasmodium of *Badhamia utricularis* and *Brefeldia maxima*. Annals of Botany, 2: 1 – 24.

Liu, P., Wang, Q., Li, Y. 2008. Study on DNA extraction and polymerase chain reaction of ITS region of rDNA from Physarum globuliferum. China-Japan Pan Asia Pacific Mycology Forum Symposium, China. 197.

Liu, P., Wang, Q., Li, Y. 2010. Spore-to-spore agar culture of the myxomycete Physarum globuliferum. Archive Microbiology, 192 (2):97 – 101.

Madelin, M. F., Audus, F., Knowles, D. 1975. Attraction of plasmodia of the Myxomycete, *Badhamia utricularis*, by Extraxts of the Basidiomycete, *Stereum hirsutum*. Journal of General Microbiology, 89: 229 – 234.

Martin, G. W., Alexopoulos, C. J. 1969. The Myxomycetes. Iowa City: University of Iowa Press, 1 – 561.

McClory, A., Coote, J. G. 1985. The chemotaxtic response of the myxomycete

Physarum polycephalum to amino acids, cyclic nucleotides and folic acid. FEMS Microbiology Letters, 26: 195 - 200.

Nakagaki, T., Yamada, H., Tóth, A. 2000. Maze-solving by an amoeboid organism. Nature, 407: 470.

Ross, I. K. 1959. Fruiting in the myxomycetes. Proceedings of Nineth International Botany Congress, Montreal, Canada. 8 - 9.

Saigusa, T., Tero, A., Nakagaki, T., et al. 2008. Amoebae anticipate periodic events. Physics Review Letters, 100: 018101 - 018104.

Singh, B. N. 1947. Studies on soil Acrasieae. 1. Distribution of species of Dictyostelium in soils of Great Britain and the effects of bacteria on their development. The Journal of General Microbiology, 28: 417 - 429.

Smart, F. 1937. Influence of certain external factors on spore germination in the myxomycetes. American Journal of Botany, 24: 145 - 159.

阿历索保罗 C. J., 明斯 C. W., 布莱克维尔 M. 2002. 菌物学概论. 姚一建, 李玉主, 译. 北京: 中国农业出版社, 769.

陈双林, 李玉. 1995. 黏菌湿室培养的初步研究. 吉林农业大学学报, 17(3):33 - 37.

谷硕, 陈小姝, 朱鹤, 等. 2011. 三种绒泡菌属原质团的培养和孢子果的诱导发生. 菌物学报, 30(4):580 - 586

谷硕. 2011. 黏菌主要种原质团的培养、生物学特性及其对细菌吞噬作用的研究. 硕士学位论文. 长春: 吉林农业大学.

姜宁. 2013. 绒泡菌目黏菌主要代表种化学成分及活性研究. 硕士学位论文. 长春: 吉林农业大学.

李晨, 王晓丽, 王晓丽, 等. 2013. 几种黏菌黏变形体形态及运动特征. 菌物学报, 32(5):913 - 918.

李新宇. 2002. 黏菌的基物培养的研究. 长春: 吉林农业大学.

李玉, 李惠中, 王琦, 等. 2007. 中国真菌志——黏菌卷(一、二). 北京: 科学出版社.

刘福杰, 潘景芝, 朱鹤, 等. 2010. 湿室培养获得秦岭地区黏菌种类. 菌物研究, 8(2): 71 - 74, 84.

刘福杰. 2010. 黏菌湿室培养及几种黏菌原质团液体培养研究. 硕士学位论文. 长春: 吉林农业大学.

刘朴, 王琦. 2006. 细弱绒泡菌的培养及个体发育初探. 菌物研究, 4(3):27 - 30.

刘朴. 2007. 主要黏菌个体发育、分子系统学及化学成分研究. 长春:吉林农业大学.

刘士德,张建华. 2004. 多头绒泡菌 PSCL 32.5 蛋白的性质及其含量在细胞周期中的变化. 遗传学报, 31(3): 305-301.

潘景芝. 2009. 吉林省不同地区基物黏菌湿室培养的初步研究. 菌物研究, 7(3-4): 142-147.

史立平,李玉. 2003. 黏菌生物学研究进展. 吉林农业大学学报, 25(1):49-53, 61.

史立平,李玉. 2004. 针箍菌的生活史. 菌物学报, 23(3):381-387.

史立平,李玉. 2005. 圈绒泡菌的生活史. 菌物学报, 24(2):292-296.

史立平,李玉. 2007. 细弱绒泡菌的生活史. 菌物学报, 26(2):211-216.

史立平,李玉. 2008. 扁绒泡菌的生活史. 菌物学报, 27(6):894-900.

史立平,王升厚,李玉. 2006. 黄柄钙皮菌的生活史. 菌物学报, 25(3): 496-501.

史立平,李玉. 2003. 不同营养介质、pH 值对黏菌孢子萌发的影响. 吉林农业大学学报, 25(3): 275-277, 281.

史立平. 2003. 绒泡菌目与团毛菌目一些黏菌生活循环的研究. 硕士学位论文. 长春:吉林农业大学.

宋晓霞. 2013. 基于生活习性、形态建成和分子特征研究黏菌四目代表类群的系统发育关系. 博士学位论文. 长春:吉林农业大学.

王琦,年淑青,李玉. 1998. 团毛菌目黏菌主要属种个体发育研究初探. 吉林农业大学学报, 20(增刊):61, 66.

王琦,年淑青,李玉. 2006. 团毛菌目黏菌个体发育比较研究. 菌物学报, 25(1):31-40.

王琦,李玉. 2006. 中国团毛菌目黏菌. 北京:科学出版社.

王晓丽,李艳双,李玉. 2007. 煤绒菌 *Fuligo septica* 显型原质团细胞核及菌核的超微结构. 菌物学报, 26(1):135-138.

王晓丽,李艳双,李玉. 2005. 几种黏菌显型原质团培养研究. 吉林农业大学学报, 27(2):140-143.

刑苗,曾宪录. 2000. 细胞松弛素 B 对多头绒泡菌有丝分裂的影响. 遗传学报, 27(1): 83-89.

徐美琴,陈萍,李玉,等. 2006. 湿室培养针叶树皮生黏菌的初步研究. 菌物研究, 4(1): 14-19.

袁海滨,陈双林. 1996. 利用湿室培养获得的 12 种黏菌. 吉林农业大学学报,

18(增刊):64-66.

赵日丰. 1983. 黏菌基物培养的研究. 吉林农业大学学报, 5(2): 61-74.

周宗璜, 张志澄, 刘宗麟. 1981. 从基物培养获得的几种黏菌. 吉林农业大学学报, 2:1-9.

周宗璜. 1981. 黏菌分类问题. 微生物学报, 8(3):129-133.

朱鹤. 2004. 团毛菌目黏菌系统分类的生物学及化学依据研究. 硕士学位论文. 长春:吉林农业大学.

Qi Wang

Doctoral supervisor of Jilin Agricultural University, Changchun, Jilin, China.

Qi Wang works at the Engineering Research Center of Chinese Ministry of Education for Edible and Medicinal Fungi, Jilin Agricultural University, China, and serves as the director of the Engineering Research Center. For now, she also takes the position of executive member of the council in Mycological Society of China. She has been the leader for a number of national and provincial projects and has published more than 70 journal articles in systematics of myxomycetes and resource development of medicinal fungus. She is also the author of five books such as *Trichiales in China*. Seven new species of myxomycetes and fungi were published and recorded in *Index of Fungi* by Qi Wang.

第三部分
Part III

生态学、生物地理学以及生物多样性
Ecology, Biogeography, and Biodiversity

Myxomycetes and Protosteloid Amoebae in the Man and Biosphere Reserve at Yangambi (D. R. Congo)

Myriam De Haan[1], Christine Cocquyt[1], George G. Ndiritu[2]

1. Botanic Garden Meise, Nieuwelaan 38, BE – 1860 Meise, Belgium;
2. National Museums of Kenya, Nairobi, Kenya

Abstract: In 2013 a field survey was undertaken in the Man and Biosphere Reserve at Yangambi (D. R. Congo, Africa) in the framework of the COBIMFO (Congo basin integrated monitoring for forest carbon mitigation and biodiversity) project, financed by the Belgian Science Policy. This reserve was deemed almost unexplored for myxomycetes, apart from four records located in the collections of the National Botanic Garden of Belgium, housed at the Botanic Garden Meise. These specimens, collected by J. Louis during 1938 – 1939 are unfortunately in a deteriorated state.

The fieldwork was performed between Oct 26 – Nov 19, 2013. A total of 280 specimens and 15 photographic records of myxomycetes were taken inside 12 forest plots, and 17 of these specimens were collected in various locations in the reserve and on its borders. In addition 63 samples were collected from 3 substrate types——aerial litter, ground litter and aerial bark——intended for agar and moist chamber cultures to record protosteloid amoebae and myxomycetes.

Preliminary results reveal about 100 species of myxomycetes identified from the field collections, as well as the moist chamber cultures and primary isolations plates. From two potential new taxa first recorded during the 2010 Congo River expedition, *Physarum* sp. collected on ground leaf litter and a *Licea*-like taxon that developed in the aerial litter cultures for protosteloid amoebae, ample material is now available to make a full to study. It seems that myxomycetes are most abundant on one particular substrate, fallen leaves of *Bellucia grossularioides*. This introduced tree, an endemic from the neo-tropics, is considered as a pioneer species in woodlands disturbed by human activities. It produces large leaves, up to 40 cm in length and 20 cm in width with a lateral curvature. The fallen leaves serve as large dome shaped moist chambers on the forest floor. Virtually all of the fruiting bodies collected in this particular micro-habitat developed on the underside of the leaves, well protected not only for the frequent heavy rainfall but also for the dehydration caused by sun exposure in between thunderstorms.

The results of the protosteloid amoebae survey are far from complete but indicate a typical species assemblage for tropical rainforests. With 80% of the aerial litter samples processed, 21 species have been recorded.

Key words:Africa; D. R. Congo; myxomycetes; protosteloid amoebae; survey

Comparative Diversity of Myxomycetes in Paleotropical (Philippines) and Temperate (USA) Forests

Thomas Edison E. Dela Cruz[1,2], **Adam W. Rollins**[3,4], **Steven L. Stephenson**[5]

1. College of Science, Research Center for the Natural and Applied Sciences, University of Santo Tomas, Manila, Philippines;

2. Fungal Biodiversity and Systematics Group, Research Center for the Natural and Applied Sciences, University of Santo Tomas, Manila, Philippines;

3. Institute of Botany and Landscape Ecology, Ernst-Mortiz-Arndt University Greifswald, Greifswald, Germany;

4. Cumberland Mountain Research Center, Lincoln Memorial University, Harrogate, Tennessee 37752, USA;

5. Department of Biological Sciences, University of Arkansas Fayetteville, Arkansas 72701, USA

Abstract: Tropical forests are generally reported to be characterized by higher diversity than temperate forests. Is this indeed the case for myxomycetes? In the present study, we compared the diversity of myxomycetes obtained from moist chamber cultures prepared aerial litter, forest floor litter and woody twigs collected from three lowland tropical forests in the Philippines and three mid-latitude temperate forests in north central Arkansas in the United States. Our results indicated that a higher value for taxonomic diversity was noted for temperate forests (SG ratio = 3.57) than for the tropical forests (SG ratio = 4.0). However, when comparing the different diversity indices, a higher value for species diversity was recorded for the Philippines (HS = 1.43) than for the United States (HS = 1.38), although this pattern contrasted with species richness (i. e., HG was 11.58 for temperate forests as opposed to 10.71 for tropical forests), albeit both types of forests had same value for evenness (E = 0.45). Among the substrata collected, the highest species diversity was noted for woody twigs, regardless of the forest type. A total of 111 species of myxomycetes were recorded in this study.

Key words: Biodiversity; slime molds; temperate myxomycetes; tropical myxomycetes

Myxomycetes in Forest Patches on Ultramafic and Volcanic Soils: Assessment of Species Diversity and Heavy Metal Biosorption

Maria Angelica D. Rea[1,3], **Nikki Heherson A. Dagamac**[3,4], **Fahrul Zaman Huyop**[5], **Roswanira A. B. Wahab**[6], **Thomas Edison E. Dela Cruz**[1,2,3]

1. The Graduate School, Research Center for the Natural and Applied Sciences, University of Santo Tomas, Manila, Philippines;
2. College of Science, Research Center for the Natural and Applied Sciences, University of Santo Tomas, Manila, Philippines;
3. Fungal Biodiversity and Systematics Group, Research Center for the Natural and Applied Sciences, University of Santo Tomas, Manila, Philippines;
4. Institute of Botany and Landscape Ecology, Ernst-Mortiz-Arndt University Greifswald, Greifswald, Germany;
5. Faculty of Biosciences and Medical Engineering;
6. Faculty of Science, Universiti Teknologi Malaysia, Johor Bahru, Malaysia

Abstract: Ultramafic and volcanic soils are exploited for industrial activities such as mining due to the high metal content of the soil. Thus, it is important that species in these areas are documented before irreversible environmental damage sets in. In this study, aerial and ground leaf litter, dead vines and twigs from forest patches on volcanic and ultramafic soils of Bataan, Pangasinan, and Zambales, Northern Philippines were placed in moist chambers and assessed for diversity and distribution of myxomycetes. From the 77% positive moist chambers for myxomycetes, a total of 33 species from 11 genera were identified. Interestingly, despite the higher heavy metal content, ultramafic forest patches had higher species diversity as compared to volcanic forest patches. Twigs from the ultramafic forest patches had also the highest number of species as compared to other substrates. In this study, eight species were abundant in both the ultramafic and volcanic forest patches, namely, *Arcyria cinerea*, *Diachea leucopodia*, *Diderma effusum*, *D. hemisphaericum*, *Didymium ochroideum*, *Perichaena chrysosperma*, *P. corticalis*, and *Physarum melleum*. Collected substrates, fruiting bodies, and plasmodia of selected myxomycetes tested for heavy metal were all positive for chromium and manganese. Interestingly, Cr and Mn contents of tested myxomycetes were equal to or higher than that of its leaf substrate. This is the first study to compare diversity and quantify Cr and Mn biosorption of myxomycetes derived from forests on ultramafic and volcanic soils.

Key words: Chromium; manganese; heavy metal; slime molds; ultramafic soil; volcanic soil

Maria Angelica D. Rea-Maminta

Instructor, University of Santo Tomas, Manila, Philippines.

Professional memberships

1. International Society for Fungal Conservation (ISFC)
2. Mycological Society of the Philippines
3. Biology Teachers Association, Inc.

Work experiences

June 2014—present: Instructor, University of Santo Tomas, Manila

April 2014—May 2014: Instructional Designer-Science, C&E Publishing, Inc., Quezon City

June 2010—March 2013: Science Coordinator, Don Bosco School-Salesian Sisters, Inc., Manila

May 2009—March 2013: Faculty, Don Bosco School-Salesian Sisters, Inc., Manila

Grants & awards

1. DOST – NSC Graduate Scholarship Grant (June 2013 – March 2014)
2. COS – Laurel Award (2008)

Major researches: 2 papers

Looking at the Diversity of Myxomycetes in the Limestone Forests of Puerto Princesa Subterranean River National Park in Palawan, Southern Philippines

Melissa H. Pecundo[2], Thomas Edison E. Dela Cruz[1,2]

1. College of Science, Research Center for the Natural and Applied Sciences, University of Santo Tomas, Manila, Philippines;
2. Fungal Biodiversity and Systematics Group, Research Center for the Natural and Applied Sciences, University of Santo Tomas, Manila, Philippines

Abstract: The Island of Palawan is considered as the last frontier of biodiversity in the Philippines. Its Puerto Princesa Subterranean River National Park has been recently awarded as one of the world's seven wonders of nature. This brings more tourists to witness its aesthetic value and enchanting beauty. However, ecotourism can also threaten an ecosystem. Documenting its biodiversity is therefore an urgent task. In this study, we assessed the diversity of myxomycetes in limestone forest habitats within Puerto Princesa Subterranean River National Park. A total of 740 moist chambers were prepared from aerial (AL) and ground (GL) leaf litter, dried inflorescences (IF), grass litter (GR), and woody twigs (TW) collected within the area. Seventy percent of the moist chambers yielded myxomycetes. Characterization of the collected specimens resulted in the identification of 33 species belonging to 16 genera and 5 taxonomic orders. Highest number of species was recorded among woody twigs (27 species) followed by aerial (21 species) and ground (15 species) leaf litter. The most abundant species included the cosmopolitan species *Arcyria cinerea* and *Stemonitis fusca*. Twenty eight species were recorded as rare. In comparison of their diversity, highest species diversity (HS = 4.99) and species richness (HG = 0.44) was again noted in woody twigs. Our study is the first extensive assessment of myxomycete diversity in the province of Palawan.

Key words: Biodiversity; karst forest; slime molds; tropical myxomycetes

Melissa H. Pecundo

Research Assistant, University of Santo Tomas, Manila, Philippines.

Educational attainment

Bachelor of Science in Biology, 2009—2013 Central Luzon State University.

Working

Fungal Biodiversity and Systematics Group, Research Center for the Natural and Applied Sciences University of Santo Tomas, Manila, Philippines.

More Additions to the Checklist of African Myxomycetes

George G. Ndiritu[1], Myriam De Haan[2]

1. Center for Biodiversity, National Museums of Kenya, P. O. Box 40658 – 00100, Nairobi, Kenya;
2. Botanic Garden Meise, Nieuwelaan 38, BE – 1860 Meise, Belgium

Abstract: During compilation of the first checklist of African myxomycetes in 2009, a number of myxomycetes records for some countries were not reported. These records were either in (i) literature not easily accessible; (ii) specimens identified but not reported or uploaded onto the Global Biodiversity Information Facility (GBIF) database or (iii) collections in museums that are not identified. To update the checklist of African myxomycetes, the first author visited Botanic Garden at Meise (Belgium) and together with the co-author scrutinized collections of myxomycetes that were collected in Africa in the past. The herbarium of the Botanic Garden in Meise holds an important collection of myxomycetes obtained from Africa in last century and was estimated to be 1094 specimens. A significant percentage of those collections belongs to Ghent University. During the present study a total of 600 specimens were checked, names verified or identified. The work was carried out in December 2012 and August 2013. The first author's stay in Belgium was supported by Belgian National Focal Point to the Global Taxonomic Initiative. Our findings show that countries with a significant number of specimens were Democratic Republic of the Congo, Rwanda and Burundi. Other countries with a substantial number were Zambia, Malawi and Morocco. Some collections were obtained in the early 1900s though most were collected between 1970 and 1990, primarily from field collections with only a few from the laboratory method of moisture chamber. Additional data will be included from recent field surveys in Kenya by the first author and in the Democratic Republic of the Congo by the co-author. Once compilation of species data is complete, it will be possible to (i) say with more certainty the number of species collected in each of the studied countries or territories in Africa, (ii) produce an updated checklist of African myxomycetes, and (iii) compile a monograph of African myxomycetes in the series of Fungus Flora of Tropical Africa published by the Botanic Garden Meise.

Key words: Africa; collections; checklist; distribution; myxomycetes; tropics

Digitalization of the Types from the N. E. Nannenga-Bremekamp Myxomycetes Collection XX

Myriam De Haan, Ann Bogaerts

Botanic Garden Meise, Nieuwelaan 38, BE – 1860 Meise, Belgium

Abstract: The myxomycetes collection of N. E. Nannenga-Bremekamp (1916 – 1996) was transferred by legacy to the Botanic Garden Meise (formerly known as the National Botanic Garden of Belgium) in 1997. This collection contains about 17,000 specimens from all parts of the world including 264 type specimens. Because of its importance this material is often used for scientific study and therefore sent on loan to different institutes worldwide. During the transport specimens can be lost or damaged, and scientific studies mostly need destructive sampling which reduces the amount of material available, certainly when only a few fruiting bodies exist. To solve these problems, herbaria have initiated digitalization projects to help the preservation of important collections and type specimens. In the case of the Nannenga-Bremekamps collection a first digitalization initiative was made by the publication of SEM photographs taken from a selection of type specimens in the series Icones Mycologiae 1982 – 1986. Only 9 types from the Nannenga-Bremekamp collection were treated, e.g. detailed descriptions were given, but only electron micrographs featured in this publication. A second digitalization project was the release of a CD – ROM in 2002 with her original drawings and notes of nearly all of the specimens, including the types. More recently in 2013 a personal initiative was started by systematically taking macro –, micro-graphs and (or) SEM images of all type specimens of myxomycetes housed in the Botanic Garden Meise.

The digitalization of a type specimen in this case means not only the imaging and databasing of the type but also the re-examing of the exsiccatum which can lead to new insights into the morphology of the concerning taxon. One example, *Licea bulbosa* Nann. – Bremek. & Y. Yamam., is presented.

Key words: Digitalization; myxomycetes

Myxomycetes Growing on Epiphytic Bryophytes: an Opportunity

Myriam De Haan

Botanic Garden Meise, Nieuwelaan 38, BE – 1860 Meise, Belgium

Abstract: Previous studies of myxomycetes growing on bryophytes have proven that only a few species can be called bryophyllous. Mosses provide an opportunity for air-born spores of myxomycetes to germinate, feed and develop in the moist conditions near the base of the thalli, and to produce fruiting bodies in the dryer environment in the top layers of the mosses.

The opportunity arose to investigate which species of myxomycetes complete their life cycle on epiphytic bryophytes in Flanders (Belgium, Western Europe). Although gaining in importance, this microhabitat has never before been systematically explored in this region. With changing climate conditions the quantity of mosses in general and more specific those growing on trees has been, and is still, steadily increasing for the last 20 years in Western Europe. The study was conducted in 16 woodlands located in four provinces of Flanders. Trunks of living trees baring epiphyticmosses were examined at 1.5 to 2 m height and specimens of myxomycetes growing among the moss thalli were collected. For comparison collections from other substrates in the same locality were also recorded. Ten species of epiphytic mosses were recorded on 12 tree species. A total of 67 specimens representing 38 taxa of myxomycetes were collected. Six of these specimens could only be identified to genus level. One of the taxa is most likely new to science. The ratio between species of mosses, trees and myxomycetes was examined.

Key words: Belgium; epiphytic bryophytes; Europe; microhabitat; myxomycetes

Some Ecological Aspects of Nivicolous Myxomycetes of the Khibiny Mts. (Kola Peninsula, Russia)

D. A. Erastova[1], Yu. K. Novozhilov[1], M. Schnittler[2]

1. V. L. Komarov Botanical Institute of the Russian Academy of Sciences, Prof. Popov St. 2, 197376 St. Petersburg, Russia;
2. Institute of Botany and Landscape Ecology, Ernst-Moritz-Arndt University Greifswald, Grimmer Str. 88, D – 17487 Greifswald, Germany

Abstract: Nivicolous myxomycetes of the Khibiny Mountains were surveyed within a vegetation gradient from the subalpine crooked-stem birch-rowan forest to dwarf shrub communities in the arctic mountain tundra during June 2012 and 2013. Both surveys yielded in totally 666 specimens representing 34 taxa (27 species, 4 varieties and 3 forms) from 8 genera and 4 families. Among them 7 are new for Russia, 30 are new for this region and 28 are new records for Fennoscandia. Most of the species (56%) were classified as rare (frequency of occurrence below 0.5% of the total number of records); only eight species were abundant (exceeding 3% of all records). Figures of the Chao 1 estimator computed from a species accumulation curve showed that our sampling effort was sufficient to recover all of the most common species in the whole studied area (80% of the expected species number; Chao 1 = 42.3 ±6.9), as well as in the subalpine crooked-stem birch forest (74%; Chao 1 = 38.1 ± 9.0). However, we did not manage to sample exhaustively the alpine mountain tundra (50%; Chao 1 = 40.3 ± 20.2). In addition, the sample coverage factor (Turing's factor) estimates the completeness of this survey as 71%, and the total number of expected species is 48. Species richness and diversity is rather high in subalpine crooked-stem birch-rowan forest (28 taxa, $H' = 2.16$) whereas alpine mountain tundra harbored a more depleted myxomycete assemblage (20 taxa, $H' = 2.02$). A comparison to the most studied nivicolous myxomycete biota of the Teberda

State Biosphere Reserve (Northwest Caucasus) based on the SØerensen similarity index Cs demonstrates a rather high similarity with myxomycete biota of the Khibiny Mts. ($Cs = 0.81$, 27 species shared). This survey is the most northern of a large-scale study of nivicolous myxomycete diversity throughout Eurasia.

Key words: Amoebozoa; arctic and alpine ecosystems; diversity; ecology; Myxogastria; slime mold; species inventory

Nivicolous Species of *Diderma* spp.: Morphology vs. Genetics

D. A. Erastova[1], Yu. K. Novozhilov[1], M. Schnittler[2]

1. V. L. Komarov Botanical Institute of the Russian Academy of Sciences, Prof. Popov St. 2, 197376 St. Petersburg, Russia;
2. Institute of Botany and Landscape Ecology, Ernst-Moritz-Arndt University Greifswald, Grimmer Str. 88, D – 17487 Greifswald, Germany

Abstract: Nivicolous species from the genus *Diderma* form a morphological complex with many transitional forms, therefore it is difficult to delimit species within this apparent morphological continuum. To elucidate their taxonomic status an in-depth morphological analysis using non-metric scaling (NMS) based on the characters of sporocarp and spore ornamentation was carried out. Morphological data were compared with the genotypes obtained from partial sequences of 18S rRNA (SSU) and EF1 – α. The collections of the focal species (*D. alpinum*, *D. fallax*, *D. globosum* var. *europaeum*, *D. meyerae*, *D. microcarpum* and *D. niveum*) were obtained from Russia (Khibiny Mts., Kola Peninsula; the Teberda State Biosphere Reserve, Northwestern Caucasus; the Valamo Island, Ladoga Lake; Vaskelovo, Leningrad oblast), Central Europe (French Alps and German Alps) and Kazakhstan (Ily-Alatau ridge, around Almaty), representing a total of 544 specimens. During this study we gained 105 sequences of the SSU gene and 57 of EF1 – α gene and compared the resulting dendrograms. A Mantel test was conducted in order to estimate the possible correlations in the different genotypes geographical distribution.

Key words: Amoebozoa; nivicolous myxomycetes; slime molds; taxonomy; molecular phylogeny

Nivicolous Myxomycetes in Agar Culture: First Results and Remaining Problems

O. N. Shepin[1], Yu. K. Novozhilov[1], M. Schnittler[2]

1. V. L. Komarov Botanical Institute of the Russian Academy of Sciences, Prof. Popov St. 2, 197376 St. Petersburg, Russia;
2. Institute of Botany and Landscape Ecology, Ernst-Moritz-Arndt University Greifswald, Grimmer Str. 88, D – 17487 Greifswald, Germany

Abstract: A total of 63 specimens of nivicolous myxomycetes representing 10 taxa (9 species and 1 variety) from 5 genera and 3 families were tested for their ability to germinate on agar plates at room temperature, using bacteria associated with the spores as food source. Germination occurred in 28 specimens representing 6 species. In some samples microplasmodia started to appear at room temperature, but in most samples further developmental stages were observed only when cultures were kept at +2°C. Two species (*Lepidoderma chailletii* and *Physarum nivale*) developed microplasmodia and even larger plasmodia, but we did not manage to induce sporulation. Partial sequences of the SSU gene obtained from cultured amoebae of two specimens from *Lepidoderma chailletii* and one from *Physarum nivale* were identical to those obtained from the spores of the specimens of origin, and comparison with GenBank sequences proved species identity. Some data about tolerance of *Lepidoderma chailletii* amoebae to low temperatures were obtained that can be helpful to understand better the ecology of nivicolous myxomycetes.

Key words: Amoebozoa; agar culture; germination; Myxogastria; nivicolous myxomycetes; temperature tolerance

Passportication for Myxomycetes Conservation

Tetiana Kryvomaz

Kyiv National Construction and Architecture University, 31, Povitroflotskyi Ave., Kyiv 03680, Ukraine

Abstract: Observation of myxomycetes conservation activity was made. The introduction of "Environmental safety passports of species" for improving of conservation policy was proposed.

First steps for conservation of myxomycetes was the creation of a myxomycetes reserve by Bruce Ing in Wales near the town of Mold, in a small town park (UK). Martin Schnittler analyzed 413 myxomycetes species from Germany by special conservation categories. Yuri Novozhilov proposed to include 21 endangered species of myxomycetes in the *Red Book of Nature of Leningradskaya oblast in Russia*, and then Alexander Lebedev recommended including in the *Red Book of Tver' oblast in Russia* 10 species of rare myxomycetes. Preliminary analyses of threat were made by author for 278 myxomycetes species of Ukraine. Species considered as endangered included 12 myxomycetes species, with 22 mainly nivicolous species being assessed as vulnerable. Detail evaluations were made for the biggest genus of myxomycetes, *Physarum*. Irina Dudka is preparing a proposal for the next edition of *Red Book of Ukraine* where myxomycetes were included. The IUCN Specialist Group promoting Conservation of Myxomycetes is beginning to prepare a foundation on which future conservation policy for Myxomycetes can be developed. The first myxomycete *Diacheopsis metallica* was published in *Red List Species on the Edge of Survival*. Evaluation of conservation status for 10 species of nivicolous myxomycetes and 10 species of order Trichiales was made. The information base included specimens, databases, bibliographic sources and field observations. Using the program "Geocat" (geocat.kew.org) estimates were made of extent of occurrence and occupancy. For each species population trend and threats were analyzed, and evaluation using IUCN criteria took place. For improving of conservation policy the introduction of

"Environmental safety passports of species" for species of myxomycetes has been proposed to make. This certificate includes data about morphology, metabolism, life cycle, geographical distribution of species and influence of abiotic and biotic factors with estimation of risks. As a result of this scientific document will be analysis of threats and conservation recommendation for myxomycetes.

Key words:Myxomycetes; conservation; passports of species

Myxomycetes Diversity in Ukrainian Forests

Tetiana Kryvomaz

Kyiv National Construction and Architecture University, 31, Povitroflotskyi Ave., Kyiv 03680, Ukraine

Abstract: The present work considers species diversity of the Myxomycetes in Ukraine, and the distribution of these organisms within the country's geobotanical regions. The study of myxomycetes distribution in Ukraine by forest types allows to identify certain patterns of taxonomic structure in different forest associations. Mixed forests in Ukraine provide the most favorable conditions and a variety of substrates for myxomycetes developing.

Key words: Myxomycetes; species diversity; forest; Ukraine

1 Introduction

Myxomycetes are eukaryotic, phagotrophic, fungus-like organisms with a plasmodial stage. In previous systems slime molds include several groups — *Myxomycetes*, *Protosteliomycetes*, *Dictyosteliomycetes*, *Copromyxida*, *Acrasida* and *Plasmodiophoromycota*. Modern systems recognize close phylogenic relationships of *Myxomycetes* ("true slime moulds") to *Protosteliomycetes* and *Dictyosteliomycetes*. The same branch *Amoebozoa* has derivation as *Copromyxida*, but it didn't find close relatives to *Acrasida* and *Plasmodiophoromycota* (Adl et al. 2012). This study is about species diversity of *Myxomycetes* in Ukraine. The first records about myxomycetes from our country were carried out in 1830 (Jundzill 1830), after scattered data appeared from time to time in disparate publications. The preliminary checklist was published only in 1996 (Minter & Dudka 1996) and special research of myxomycetes begins mainly on protected territory (Dudka et al. 1999; Krivomaz 2003, 2004; Leontyev 2006). Ukraine has big forest and forest-steppe zones, where moisture and decaying organic matter provides favorable substrates for myxomycetes. The goal of this research is determination of myxomycetes diversity and features of their distribution in different forest types in Ukraine.

2 Methods and materials

This research of myxomycetes diversity was carried out in forest and steppe-forest zones of Ukraine (Fig. 1). It was held from 1994 to 2009 in 18 (from totally 26) geobotanical regions of Ukraine: Carpathian Forests, Donetskiy Cerea-Meadow Steppe, Left-bank Cerea-Meadow Steppe, Left-bank Forest-Steppe, Left-bank Polissya, Little Polissya, Prykarpattya Forests, Right-bank Cerea Steppe, Right-bank Cerea-Meadow Steppe, Right-bank Forest-Steppe, Right-bank Polissya, Roztochchya Forests, South Bank of Crimea, Starobilsky Cerea-Meadow Steppe, Volinskiy Forest-Steppe, Western Forest-Steppe, Western Polissya, Western Ukrainian Forest, Zacarpathian Forest. Collecting was made by Tetyana Krivomaz (solid marks on map) and Irina Dudka (hollow marks on map). The special attention was paid to protected territories and 20 protected areas of Ukraine were studied during research. In this study fields material were analyzed from Dniprosko-Orilsky, Kaniv, Crimea, Opuk, Polissya, Rivne, (Roztochya), Cheremsky Nature Reserves; Yalta Mountain-forest Nature Reserve; Carpathian Biosphere Reserve; Carpathians, Desna-Starogutsky, Ichnya, Holosievo,(The Holy Mountain), Mezyn,(PodilskiTovtry),(Synevir), Uzhanskiy and Shatsky National Parks. Collecting of myxomycetes also was made on not-protected

Fig. 1 The places of Myxomycetes collecting in Ukrainian forests zones.

areas in Crimea, Zhitomir, Kiev, Kirovograd, Lviv, Kherson and Chernigov regions of Ukraine. In total 1643 myxomycetes samples were collected using standard field collection and laboratory identification methods(Ing 1999; Leontyev 2008; Stephenson et al. 1993). The herbarium data from Ukrainian collection of Bruce Ing, Dmitry Leontyev and Katerina Romanenko were counted. For general analysis of myxomycetes conservation characteristics also literature data were used.

3 Results

As a result of the research, 160 species were identified belonging to 34 genera from 12 families and 5 orders of the class *Myxomycetes*. For the territory of Ukraine 33 of those species were new: *Collariarubens*, *Comatrichasuksdorfii*, *C. tenerrima*, *Crateriumconcinnum*, *Didermaalpinum*, *D. chondrioderma*, *Echinosteliumbrooksii*, *E. corynophorum*, *Hemitrichiaintorta*, *Lamprodermacribrarioides*, *L. echinosporum*, *L. ovoideoechinulatum*, *L. ovoideum*, *L. pulveratum*, *L. spinulosporum*, *L. splendens*, *L. zonatum*, *Lepidodermacarestianum*, *L. chailletii*, *Liceadenudescens*, *L. marginata*, *L. pedicellata*, *Oligonemaaurantium*, *Physarumalbescens*, *Ph. alpestre*, *Ph. auriscalpium*, *Ph. bethelii*, *Ph. pulcherripes*, *Ph. tenerum*, *Stemonitismussooriensis*, *Stemonitopsisamoena*, *S. gracilis*, *Trichiaalpina*. For 18 Ukrainian geobotanical regions 142 species were first recorded, and 131 were new for 20 protected areas of Ukraine. All these data represent over 57% of the total known Ukrainian myxomycetes biota. After this study 278 species of 43 genera, 13 families and 5 orders are known for the territory of Ukraine in total together with previous research and literature data. All these compose just over 30% of the world's total diversity for this group (Fig. 2).

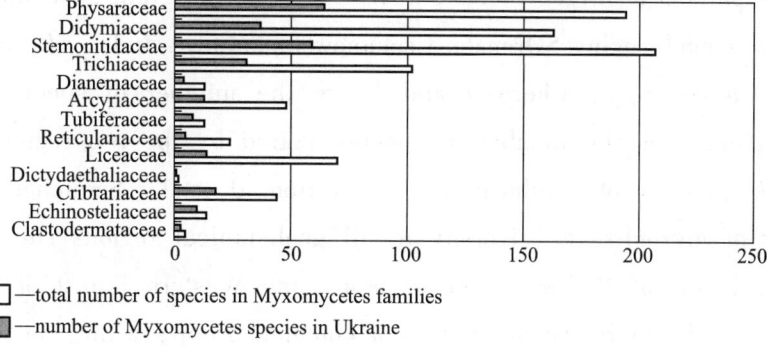

Fig. 2 Comparing of the Ukraine biota with the total number of known Myxomycetes species by families.

All species were classified according to their abundance: species representing more than 3% of the total number of collections were considered as abundant, those falling between 1.5% – 3% as common, between 0.5% and up to 1.5% as occasional and those less than 0.5% as rare (Stephenson et al. 1993). According to this classification 278 species are apportioned as follows: 31 abundant, 49 common, 49 occasional and 149 rare for country. Such a large number of "rare" species are evidence of general deficiency study of myxomycetes in Ukraine. The preliminary analysis identified only 33 species as rare for Ukraine in comparison with their world distribution: *Arcyriaglobosa*, *Badhamiamelanospora*, *Clastodermadebarianum*, *Comatrichaellae*, *C. longipila*, *Cribrariamacrocarpa*, *C. splendens*, *Didermachondrioderma*, *D. cingulatum*, *D. montanum*, *Didymium sturgisii*, *Echinosteliumapitectum*, *Fuligomuscorum*, *Hemitrichiaintorta*, *Lepidodermatigrinum*, *Liceainconspicua*, *L. tenera*, *Oligonemaflavidum*, *Perichaenapedata*, *Physarumcitrinum*, *Ph. confertum*, *Ph. conglomeratum*, *Ph. decipiens*, *Ph. digitatum*, *Ph. gyrosum*, *Ph. licheniforme*, *Ph. murinum*, *Ph. notabile*, *Ph. oblatum*, *Stemonaria longa*, *Stemonitopsisamoena*, *S. gracilis*, *Trichialutescens*. By comparing of list of myxomycetes from 3 most studied geobotanical regions (Carpathian Forest, KharkivForest-steppe, Mountain Crimea) 74 species can be considered as nucleus of myxomycetes biota in Ukraine. All these regions are very different by climate, relief and vegetation, so this nucleus suggests for any forest ecosystem in Ukraine. The most abundant myxomycetes species in Ukraine are *Arcyriacinerea*, *A. ferruginea*, *A. pomiformis*, *Comatrichanigra*, *Cribrariacancellata*, *Fuligoseptica*, *Hemitrichiaclavata*, *Lycogalaepidendrum*, *Stemonitisaxifera*, *S. fusca*, *Stemonitopsistyphina*, *Trichiafavoginea*, *T. varia*, *Tubiferaarachnoidea*.

The comparison of myxomycetes species lists from different geobotanical regions of Ukraine was made using SØrensen-Czekanowskisimilarity index: $K_{sc} = 2a / [(a + b) + (a + c)]$, where a and b are the number of species in regions respectively, and c is the number of species shared by the two regions (Leontyev 2008). The quotient of similarity (K_{sc}) is from 0 to 1. Sequence of pairwise comparisons of myxomycetes diversity in all geobotanical regions showed that the myxomycetes biotas of Polissya, Forest-steppe and Western Forest of Ukraine are rather similar and can be considered as a common group, while the myxomycetes biotas of the Carpathians, Crimea and the Steppe zone more separate (Fig. 3). This comparison mainly shows degree of myxomycetes scrutiny in compared areas, but it clearly attests that taxonomic structure of myxomycetes biota depends on main trees species in forest ecosystems.

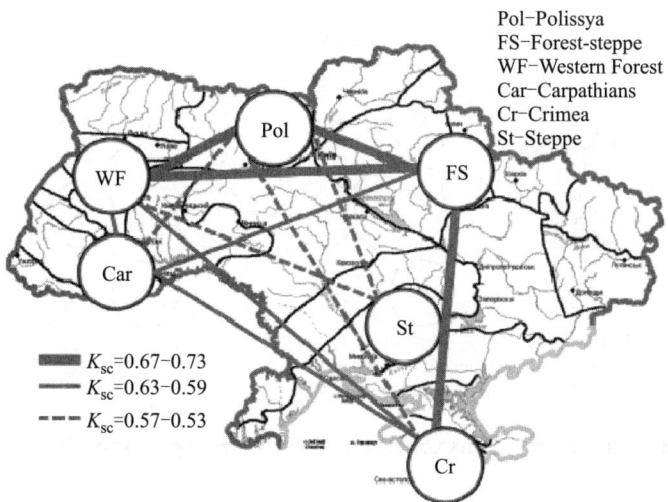

Fig. 3 **Myxomycetes diversity in different geobotanical regions of Ukraine.**

Myxomycetes species compositions change in accordance with variation of vegetation. The spread of *Fagussylvatica* is limited to the western regions of Ukraine and respectively species numbers of family Physaraceae increase in the western forests. *Pinussylvestris* formations dominate the north; therefore members of families Stemonitidaceae and Cribrariaceae rise in the northern regions of Ukraine. *Carpinusbetulus* forests are concentrated on the right bank of the country and species from family Trichiaceae was found more compared with other regions of Ukraine. The forest component of *Quercusrobur* decreases towards the west to the east and the number of family Didymiaceae increases there.

5 Discussion

Myxomycetes species diversity mainly depends on vascular plants providing various types of substrates. The study of the myxomycetes distribution in Ukraine by forest types allows to identify certain patterns of taxonomic structure in different forest associations. It is possible to predict myxomycetes species compositions depending on the micro-climatic conditions and availability of appropriate substrates. In coniferous forests made by spruce pine (Pinaceae) almost equal myxomycetes species composition were identified, except family Didymiaceae. In alder and hornbeam forests (Betulaceae) myxomycetes of families Trichiaceae, Stemonitidaceae,

Arcyriaceae and Physaraceae dominate. In oak and beech forests (Fagaceae) species from families Stemonitidaceae and Physaraceae prevailed (Fig. 4).

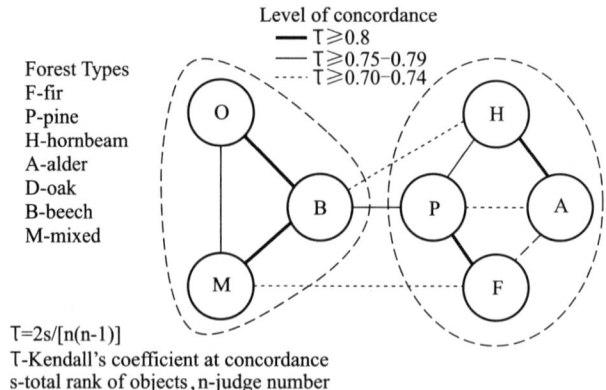

Fig. 4 The similarity of myxomycetes species compositions in forest types.

In general family Trichiaceae common in hornbeam forests, Cribrariaceae and Stemonitidaceae abundant in conifers, Arcyriaceae has a wide range of adaptation to different forest types. Only five myxomycetes species (*Arcyriaferruginea*, *A. obvelata*, *Lycogalaepidendrum*, *Stemonitisaxifera* and *S. fusca*) were common in all researched forest communities of Ukraine. This study shows that mixed forests in Ukraine provide the most favorable conditions and a variety of substrates for myxomycetes developing.

References

Adl, S. M., Simpson, A. G. B., Lane, C. E. et al. 2012. The revised classification of eucaryotes. Journal of Eukaryotic Microbiology, 59(5): 429 – 514.

Dudka, I. O., Kuzub, V. V., Romanenko, E. A. 1999. Myxomycetes of the Yalta mountain-forest nature reserve (Ukraine, Crimea). Mycology and Phytopathology, 33 (5): 307 – 315 [in Russian].

Ing, B. 1999. The Myxomycetes of Britain and Ireland. An identification Handbook. Slough: The Richmond Publishing Co. Ltd. 374.

Jundzill, J. 1830. A description of wild and domesticated plants in Lithuania, Volhynia, Podolia and Ukraine. Lithuania, Vilnius. 554 – 556.

Krivomaz, T. I. 2003. Myxomycetes of Rivnens'kiy nature reserve. Ukrainian

Botanical Journal, 60 (6): 633 – 642 [in Ukrainian].

Krivomaz, T. I. 2004. Myxomycetes of Shats'kiy National park. Ukrainian Botanical Journal, 61 (5): 45 – 53 [in Ukrainian].

Leontyev, D. V. 2008. Floristic analysis in mycology. Kharkiv, 110. [in Russian].

Leontyev, D. V. 2006. Myxomycetes (Myxomycota) species composition of National Park "Gomolsha forests" (Ukraine). Mycology and Phytopathology, 40 (2): 101 – 107 [in Russian].

Minter, D. W., Dudka, I. O. 1996. Fungi of Ukraine: a preliminary checklist. Egham, Surrey, UK, International Mycological Institute & Kiev, Ukraine, M. G. Kholodny Institute of Botany. 361.

Stephenson, S. L., Kalyanasundaram, I., Lakhanpal, T. N. 1993. A comparative biogeographical study of myxomycetes in the mid-Appalachians of eastern North America and two regions of India. Journal of Biogeography, 20: 645 – 657.

Tetiana Kryvomaz
Qualifications
Ph D, mycology; BSc in botany, T. Shevchenko National University of Kiev; Polytechnic Institute of Kiev, chemistry.
Current position
Senior lecturer of Kyiv National Construction and Architecture University; journalist, subeditor of the journal *Pharmacevt practic*; leader of NGO "Ukrainian Ecological Society."
Publication
72 scientific works and 50 popular-science articles.
Projects
13 projects.
Expeditions
13 times.

Higher Myxomycete Diversity in Mountainous Vegetation than Agricultural Plantation?
—An Evidence from Mt. Kanlaon National Park, Negros Occidental, Philippines

Julius Raynard Alfaro[1], **Donn Lorenz Alcayde**[1], **Joel Agbulos**[1], **Nikki Heherson Dagamac**[3], **Thomas Edison E. Dela Cruz**[1,2]

1. College of Science, Research Center for the Natural and Applied Sciences, University of Santo Tomas, Manila, Philippines;
2. Fungal Biodiversity and Systematics Group, Research Center for the Natural and Applied Sciences, University of Santo Tomas, Manila, Philippines;
3. Institute of Botany and Landscape Ecology, Ernst-Mortiz-Arndt University Greifswald, Greifswald, Germany

Abstract: Higher floral and faunal biodiversity is expected in multi-species-covered mountainous forests than in mono-typic agricultural plantations. Is this also true for slime molds? This study comparatively evaluated the occurrence, diversity, and community assemblages of myxomycetes between agricultural and mountain forest areas, here represented by sugar cane plantations and Mt. Kanlaon, respectively. A total of 23 species of myxomycetes were collected and identified in Mt. Kanlaon National Park in Negros Occidental, Central Philippines. Morphological characterization identified these as belonging to the genera *Arcyria*, *Ceratiomyxa*, *Collaria*, *Craeterium*, *Cribraria*, *Diderma*, *Didymium*, *Hemitrichia*, *Lamproderma*, *Physarum*, *Stemonitis*, and *Trichia*. In contrast, only one species of myxomycetes, i.e. *Arcyria cinerea*, was recorded in the sugar cane plantations, indicating that mountain forests have a higher taxonomic and species diversity. This research is the first study to report the myxomycetes from Negros Occidental.

Key words: Agricultural plantation; biodiversity assessment; mountain forests; slime molds; species list

Thomas Edison E. Dela Cruz
Professor of Microbiology, Chair, Department of Biological Sciences Fellow, Philippine Academy of Microbiology, Philippines.

Highest educational attainment
Doctor of Natural Sciences (Dr. Rer. Nat.), Institute of Microbiology, Faculty of Life Sciences Technical University of Braunschweig, Braunschweig, Germany.

Work experiences
Professorial Lecturer, Graduate School, University of Santo Tomas, Manila, Philippines.

Selected publications
12 papers.

Papers presented
A. Invited Lectures
International Conferences and Scientific Meetings: 9
Local Conferences, Scientific Meetings and University Activities: 21
B. Oral Paper Presentation
International Conferences and Scientific Meetings: 23
Local Conferences, Scientific Meetings and University Activities: 31
C. Poster Presentation
International Conferences and Scientific Meetings: 17
Local Conferences, Scientific Meetings and University Activities: 57
D. Workshops Conducted/Organized (as Organizer and/or Resource Person):15

A Look at the Diversity of Myxomycetes in the Mountain and Coastal Forests of Puerto Galera, Oriental Mindoro

Nathan S. Batungbacal[1], **Carmela Rina T. Bulang**[1], **Akira Gioia R. Cayago**[1], **Soohyun Jung**[1], **Nikki Heherson A. Dagamac**[3], **Thomas Edison E. Dela Cruz**[1,2]

1. College of Science, Research Center for the Natural and Applied Sciences, University of Santo Tomas, Manila, Philippines;
2. Fungal Biodiversity and Systematics Group, Research Center for the Natural and Applied Sciences, University of Santo Tomas, Manila, Philippines;
3. Institute of Botany and Landscape Ecology, Ernst-Mortiz-Arndt University Greifswald, Greifswald, Germany

Abstract: Myxomycetes are commonly associated with decaying plant materials in all types of terrestrial ecosystems. In the Philippines, these organisms have been reported in island and lowland forests. However, many areas in the country remain unexplored for myxomycetes. In this research study, the diversity of myxomycetes in the coastal, community and mountain forests of Puerto Galera in Oriental Mindoro was assessed and compared. Field specimens were collected directly from the forest sites while moist chambers were set up from twigs, dead vines, and ground and aerial leaf litter. A total of 42 species were recorded in the study. They belong to 15 genera, namely *Arcyria*, *Ceratiomyxa*, *Collaria*, *Comatricha*, *Cribraria*, *Didymium*, *Diachea*, *Diderma*, *Echinostelium* *Hemitrichia*, *Lamproderma*, *Lycogala*, *Perichaena*, *Physarum*, and *Stemonitis*. Mountain and coastal forests had a higher number of species than community forest. Mountain and coastal forests also had the same species diversity index while a lower value was reported for community forest.

Arcyria cinerea was found to be the most abundant among the myxomycetes recorded. This study is the first extensive report on myxomycetes in Puerto Galera and in the island of Mindoro.

Key words: Species occurrence; slime molds; species listing; tropical forests

Myxomycete Diversity and Ecology in Tropical Forests of Southern Vietnam: First Results and Perspectives

Yu. K. Novozhilov[1], Yu. A. Morozova[2], A. V. Alexandrova[3], E. S. Popov[1], A. N. Kuznetzov[4]

1. V. L. Komarov Botanical Institute of the Russian Academy of Sciences, Prof. Popov St. 2, 197376, St. Petersburg, Russia;
2. St. Petersburg State University, Faculty of Biology, 7 – 9, Universitetskaya nab., St. Petersburg, 199034, Russia;
3. Lomonosov Moscow State University, Faculty of Biology, Leninskie Gory 1 – 12, 119991, Moscow, Russia;
4. Joint Russian-Vietnamese Science and Technological Tropical Center, Hanoi, Vietnam

Abstract: Myxomycete assemblages were surveyed during October – December 2010 – 2013 in lowland tropical forests of Cat Tien National Park (CTNP, 11°21′ – 11°48′N, 107°10′ – 107°34′E) and Vinh Cuu Nature Reserve (VCNR, 11° – 11°30′N, 106°54′ – 107°13′E) and in mixed montane tropical forests and tropical cloud forests of Bi Dup-Nui Ba Nature Reserve (BDNB), centered in the Bi Dup Mountain massive (12°08′N, 108°40′E), belonging to the Da Lat Plateau (Lam Dong Province) of southern Vietnam. We present data on biodiversity of myxomycetes in southern Vietnam with special emphasis on vegetation types and the associated substrates, and a comparison of our data with results from other surveys in tropical Southeast Asia. Specific objectives were (ⅰ) to characterize the assemblage of commonly occurring species by assessing their abundances and (ⅱ) to obtain data on the distribution of myxomycetes along an elevational gradient. This study is based on both field collections and those from moist chamber cultures prepared with ground litter, aerial litter and bark of living trees and liana. Within a vegetation gradient reaching from deciduous monsoon tropical lowland forests in CTNP and VCNR (elevations 70 – 85 m a. s. l.) to mixed montane tropical forests and coniferous mountain tropical forests (elevations 1400 – 1600 m) and cloudy tropical forests (elevations 1650 – 1750 m) in BDNB, myxomycete assemblages were surveyed. A total of 1477 records representing 135 species of myxomycetes in 22 genera were considered; including 494 field records (79 species) and 982 records (94 species) observed in 1513 moist chamber cultures of various decaying plant material. The majority of species (110) were classified as rare (frequency of occurrence below 0.5% of the total of 1477 records); only nine species were found to be abundant (exceeding 3% of all records). We report 134 species the first time for Vietnam, and three of them are species new to science. An evaluation of these data by a species accumulation curve estimated that between 75% to 83% of the species richness had been recorded. Shannon diversity and species richness reached maximum values in the deciduous monsoon tropical lowland forests, whereas coniferous mountain tropical forests and tropical cloud forests had the most depauperate but most specific myxomycete assemblages. The assemblages associated with ground and aerial litter are most diverse, but the one associated with bark of coniferous trees is the most distinctive. Assemblages associated with the monsoon tropical forests of the southeastern Vietnam displayed a high level of similarity to those of other tropical regions for which data exist.

Key words: Amoebozoa; diversity; ecology; Myxogastria; slime mold; species inventory; Southeast Asia; tropical forest

Four Years in the Caucasus: Observations on the Ecology of Nivicolous Myxomycetes

Martin Schnittler[1], Daria A. Erastova[2], Oleg N. Shchepin[2], Eva Heinrich[1], Yuri K. Novozhilov[2]

1. Institute of Botany and Landscape Ecology, Ernst Moritz Arndt University Greifswald, Soldmannstr. 15, D – 17487 Greifswald, Germany;
2. V. L. Komarov Botanical Institute of the Russian Academy of Sciences, Prof. Popov St. 2, 197376 St. Petersburg, Russia

Abstract: We report the results of four years' survey of the abundance and habitat requirements of nivicolous myxomycetes at the northwestern Greater Caucasus ridge. An elevational transect spanning 3.66 km from 1700 to 3000 m a.s.l. was established at the summit Malaya Khatipara situated within the Teberda State Biosphere reserve. Revisiting this transect every year between 2010 and 2013, we recorded 1177 fructifications of nivicolous myxomycetes, with 700 of these determined to 44 taxa. Fructifications developed usually at the margin of a snow field or in its close vicinity. Abundance of myxomycetes varied extremely between years, ranging from nearly zero to hundreds of fructifications. At sites with known myxomycete occurrences we established 16 data loggers in the years 2011 and 2012, measuring relative humidity and temperature at the soil surface. Together with weather data recorded on the nearby Klukhor pass, these data explain the observed extreme fluctuations in myxomycete abundance. If the uppermost soil layer freezes before snow cover, freezing temperatures may be preserved until snow melts, which is indicated by a phase of constant temperature (0℃). If snow falls before frost, the insulating features of the snow allow for temperatures around 0℃ for most of the winter. In addition, both weather data and the data logger revealed a deep frost before the first snow falls in 2011. Temperature minima exceeding – 10℃ at the soil surface most likely caused a nearly complete failure of myxomycete fruiting at the next spring. Although the overall snow cover was much lower in 2012, frost occurred only after the first snow falls, and myxomycetes fruited again at higher elevations. Our data show that frost(snow) events in the previous autumn are critical for the temperature regime under the snow, determining together with the absolute duration of snow cover the time period suitable for amoebal growth. At sunny days, a steep temperature gradient in soil temperatures develops with the retreating snow, which may be the reason that myxomycetes fruit often immediately after snow melt. These assumptions are supported by amoebal cultures germinated from spores, which grew well over several months at temperatures between 1℃ and 4℃ but did not survive frost below −10 ℃.

Key words: Amoebal growth; data logger; frost; undersnow microbial communities

Myxomycetes of Vyatka River Valley

V. A. Sysuev, A. A. Shirokikh, I. G. Shirokikh

N. V. Rudnitski Zonal North-East Agricultural Research Institute, Kirov 610007, Russia

Abstract: Results of researches of a myxomycetes specific variety in natural and urban ecosystems of a Vyatka river basin in a subzone of a southern taiga of the European North-East are generalized. Researches are conducted by a route method, as well as by the method of moist chambers. The revealed species of myxomycetes were photographed under field conditions with Canon EOS 5D Mark II having adaptations for macro shooting. Identification was done by means of microscope Leica DM 2500, according to Yu. K. Novozhilov's (1993) qualifier, G. Nojbert's 3 - volume guide (1993, 1995, 2000) and the Internet resource http://www.discoverlife.org/.

In a large forest located in Kirov suburb, 14 species of myxomycetes, concerning 5 orders and 5 families were revealed. The majority of the found out species are cosmopolitans, widespread in the territory of Russia. The species found out in a large forest belongs to families *Physaraceae*: *Physarum viride* (Bull.) Pers., *Ph. nutans* Pers., *Badhamia macrocarpa* (Ces.) Rost., *Fuligo septica* (L.) Wigg., *Leocarpus fragilis* (Dicks.) Rost.; *Arcyriaceae*: *Arcyria denudata* (L.) Wettst., *A. cinerea* (Bull.) Pers., *A. pomiformis* (Leers) Rost., and *Trichiaceae*: *Trichia decipiens* (Pers.) Macbr., *T. favoginea* (Batsch) Pers., *T. varia* (Pers ex J. F. Gmel) Pers. Along with listed, species *Stemonitis fusca* Roth. and *Lycogala epidendrum* (L.) Fr. were constantly found out on foozles and trunks of tumbled down trees. In the summer, after rains, on trunks of the tumbled down trees and foozles, plentiful growth of *Ceratiomyxa fruticulosa* (Mull) Macbr. was observed. Higher, than in the summer, a specific variety of myxomycetes is noted in the fall.

A specific variety of myxomycetes in parks of a Kirov city has appeared essentially lower, than in suburban wood phytocenoses. It was possible to find out representatives of two species only: *L. epidendrum* (L.) Fr. (order *Liceales*) and *Mucilago crustacea* F. H. Wigg (order *Physarales*). Both species are noted on foozles.

In the route researches spent in woods of the State natural reserve " Nurgush " (southwest suburb of Srednevjatsky lowland on the average of a watercourse of Vyatka river) 20 species of myxomycetes, concerning 5 orders and 7 families are revealed. Xylobiontic substrate complex was characterized by the greatest variety—85 % of all found out species concerning mainly to families *Arcyriaceae* (23.5%) and *Physaraceae* (23.5%).

Nine species of myxomycetes are revealed in epiphyte and covering substrate complexes of the territory of the Reserve. Representatives of families *Physaraceae* (44.4% of total number of species), *Stemonitidaceae* (22.2%), *Trichiaceae* (11.2%), *Arcyriaceae* (11.2%), and *Didymiaceae* (11.0%) dominated. The wide circulation of the species belonging to family *Physaraceae* in all substrate complexes indicates ecological plasticity of many representatives of this taxon.

About 160 samples of rotten wood and a bark of the trees collected in the territory of the Reserve had been analyzed by a method of moist chambers. As a result 13 additional species have been revealed which have not been found out during route researches.

As a whole 33 species of myxomycetes of various substrate complexes (epiphytic, xylobiontic, and covering) are revealed in the surveyed especially protected territory of Reserve " Nurgush. "

Key words: Myxomycetes; southern taiga subzone; European North-East; species diversity; substrate complex; natural ecosystem; urban ecosystem

1 Introduction

In different geographical areas of the Russian Federation, myxomycetes were studied very unevenly, with a total of about 310 species observed (Novozhylov et al. 2005). The Kirov region in different years has been only sporadic attempts to study slime molds (Khizhnyakova & Ronko, 2009; Shirokikh & Shirokikh, 2010; Shirokikh, 2011). We have investigated diversity of myxomycetes species in three habitats: suburban mixed forest, Kirov city parks, and reserve "Nurgush" (Fig. 1).

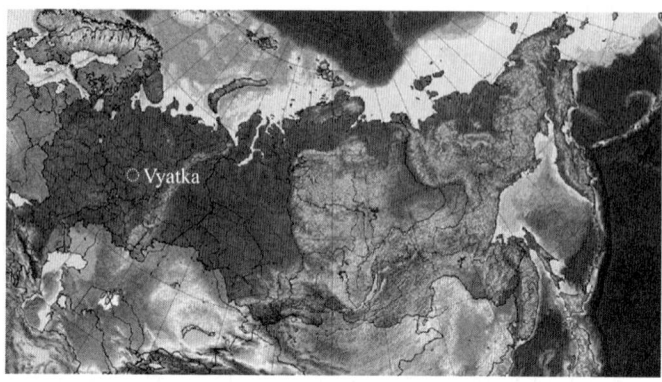

Fig. 1 The research location.

2 Method

To study the species diversity of myxomycetes, the route method was used. Throughout selected routes, we examined trunks of fallen trees, rotten stumps, old fruiting bodies of wood-decaying fungi and litter. Discovered species of myxomycetes were photographed in the field using camera Canon EOS 5D Mark II and devices for macro. Species of myxomycetes were identified by standard technique using a microscope Leica DM 2500, handbook of Y. K. Novozhylov (1993), 3 – volume guide of G. Neubert et al. (1993, 1995, 2000) and http://www.discoverlife.org/.

3 Results

In a large forest located in Kirov suburb, 14 species of myxomycetes, concerning 5 orders and 5 families were revealed. The majority of the found out species are cosmopolitans widespread in territory of Russia. The species found out in a large forest belongs to families *Physariaceae*: *Physarum viride* (Bull.) Pers., *Ph. nutans*

Pers. , *Badhamia macrocarpa* (Ces.) Rost. , *Fuligo septica* (L.) Wigg. , *Leocarpus fragilis* (Dicks.) Rost. ;*Arcyriaceae*; *Arcyria denudata* (L.) Wettst. (Fig. 2) , *A. cinerea* (Bull.) Pers. , *A. pomiformis* (Leers) Rost, and *Trichiaceae*: *Trichia decipiens* (Pers.) Macbr. (Fig. 3) , *T. favoginea* (Batsch) Pers. , *T. varia*(Pers ex J. F. Gmel) Pers. Along with listed, species *Stemonitis fusca* Roth. and *Lycogala epidendrum* (L.) Fr. were constantly found out on foozles and trunks of tumbled down trees. In the summer, after rains, on trunks of the tumbled down trees and foozles, plentiful growth of *Ceratiomyxa fruticulosa* (Mull) Macbr. (Fig. 4) was observed. Higher, than in the summer, a specific variety of myxomycetes is noted in the autumn.

Fig. 2　*Arcyria denudata* (L.) **Wettst.**

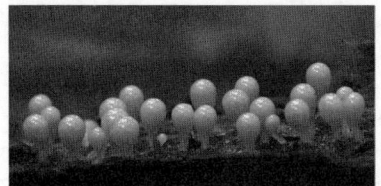

Fig. 3　*Trichia* **decipiens** (Pers.) **Macbr.**

Fig. 4　*Ceratiomyxa fruticulosa* (Mull) **Macbr.**

A specific variety of myxomycetes in parks of Kirov city has appeared essentially lower, than wood phytocenoses. It was possible to find out representatives of two species only: *L. epidendrum* (L.) Fr. (order *Liceales*) and *Mucilago crustacea* F. H. Wigg (order *Physarales*). Both species are noted on foozles.

In the route researches spent in woods of the State natural reserve "Nurgush" (southwest suburb of Srednevyatsky lowland on the average of a watercourse of Vyatka river) 20 species of myxomycetes concerning 5 orders 7 families are revealed. Xylobiotic substrate complex was characterized to be of the greatest variety——85% of all found out species concerning mainly to families *Arcyriaceae* and *Physaraceae*. Nine species of myxomycetes are revealed in epiphyte and litter substrate complex of territory of the Reserve. Representatives of families *Physaraceae* (44.4% of total number of species), *Stemonitidaceae*, *Trichiaceae*, *Arcyriaceae*, and *Didymiaceae* dominated.

About 160 samples of rotten wood and bark of the trees collected in territory of Reserve had been analyzed by method of moist chambers. As a result 13 additional species have been revealed which have not been found out during route researches.

As a whole, 33 species of myxomycetes of various substrate complex (epiphytic, xylobiotic, and litter) are revealed in the surveyed especially protected territory of Reserve "Nurgush" (Fig. 5).

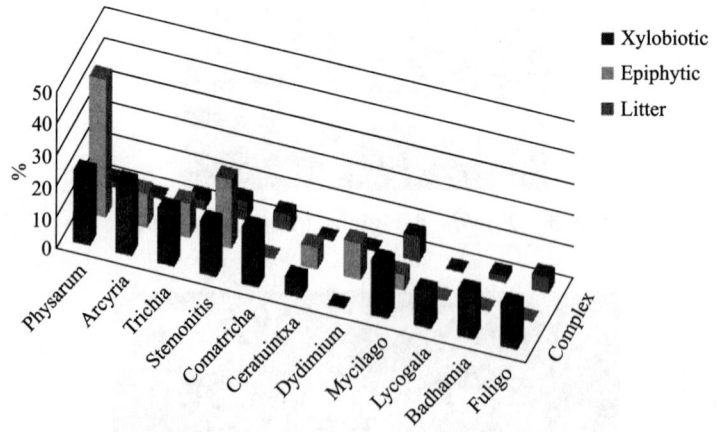

Fig. 5　The ratio of the genera of slime molds in the substrate complex.

4　Discussion

Species diversity of slime molds in phytocenosis is primarily due to their physiology and morphology. Substrate, temperature and humidity are the most important abiotic factors that influence the spread of certain substrate complexes of slime molds. The large number of fallen trees, old stumps and powerful moss cover in the woodland forest provides commuter widespread xylobiotic myxomycetes forming large sporophores. Surveyed by shuttle forest townspeople used as a recreation area and subjected to high anthropogenic pressure, but despite this, the diversity of species of slime molds in this phytocenosis remains high. At the same time, the diversity of species of myxomycetes in city parks was very low. Obviously, reduction of species diversity of myxomycetes is not only due to pollution of the urban environment, but also because of the lack of suitable substrates for their reproduction.

5 Conclusion

The study of species diversity of myxomycetes in the surveyed areas of Valley Vyatka identified 37 species. Main diversity of species of slime molds was found in forest ecosystems. Significant species richness of different slime molds was found in specially protected area reserve "Nurgush." In urban parks myxomycetes presented singly, owing to the lack of suitable substrates for their development and, possibly, the high concentration of pollutants.

References

Khizhnyakova, A. S., Ronko, R. V. 2009. Myxomycetes of reserve "Nurgush" // Research as a basis for the protection of natural complexes reserves and sanctuaries. Proceedings of the All-Russian Scientific and Practical Conference. Kirov, 159 – 160.

Neubert, H., Nowotny, W., Bauman, K. 1993. Die Myxomyceten// 3 Banden. B. 1 Ceratiomyxales, Liceales, Echinosteliales, Trichiales. Berlin: Karlheinz Bauman Verlag Comaringen. 343s.

Neubert, H., Nowotny, W., Bauman, K., et al. 1995. Die Myxomyceten// B. 2 Physarales. Berlin: Karlheinz Bauman Verlag Comaringen. 368s.

Neubert, H., Nowotny, W., Bauman, K., et al. 2000. Die Myxomyceten// B. 3 Stemonitales. Berlin: Karlheinz Bauman Verlag Comaringen. 392s.

Novozhylov, Y. K., Zemlianskaia, I. V., Schnittler, M. 2005. Corticolous myxomycetes in desert of the northwestern Caspian lowland. Mycology and Phytopatology, 39(5): 43 –54.

Novozhylov, Y. K. 1993. Definitorium Fungorum Rossiae. Divisio Myxomycota. Classis Myxomycetes. Petropolis: "Nauka." 288.

Shirokikh, A. A., Shirokikh, I. G. 2010. Diversity of myxomycetes in a forested area of Kirov // Immunopathology, Allergology, Infektology. N 1. 54 –55.

Shirokikh, A. A. 2011. Xylobiotic myxomycetes of reserve "Nurgush"// Proceedings of the National Nature Reserve "Nurgush." Kirov. Vol. 1. 182 –187.

Alexandr Shirokikh

Professor of N. V. Rudnitski Zonal North – East Agricultural Research Institute, Russia

Education Experiences

I work at the Vyatka State University of Humanities as Professor of the Department of Ecology and on the branch of St. Petersburg Institute of Foreign Economic relations, Economics and Law as Professor of the Department of General Humanitarian and natural – science disciplines. I have 10 years experience.

Work Experiences

I worked at the Meadow, Bog and Grassland Experimental Station of Kirov, V. R. Williams All – Russian Institute of Forages as Junior Researcher, Senior Researcher, (1983 –1994).

I work presently at the N. V. Rudnitski Zonal North – East Agricultural Research Institute as Leading Researcher of the laboratory of Plants and Microorganisms Biotechnology (1994 – at is currently).

I am member of the Scientific and Technical Council of the State Nature Reserve "Nurgush", member of the National Academy of Mycology Russia.

Myxomycete Diversity and Distribution in the Mountain Valley of Kamikochi in the Northern Japan Alps

Kazunari Takahashi[1], Yuichi Harakon[2]

1. Okayama University of Science High School, 1 – 1 Ridai-cho, Kitaku, Okayama City, Okayama 700 – 0005, Japan;
2. Department of Forest Product Science, Kyushu University, Tsubakuro 394, Sasaguri, Kasuya, Fukuoka 811 – 2415, Japan

Abstract: Most myxomycetes are distributed in coarse woody debris in forests and play an important role in the detritus food chain and in the process of material recycling. However, the species diversity and the ecological characteristics in a primitive forest are little known in Japan. The present study investigated myxomycete species diversity and distribution on logs in a natural conservation forest of the Kamikochi that is located in Central Mountain National Park of Japan, situated in a subalpine great valley, altitude 1500 – 1600 m, of the Northern Japan Alps, Central Japan.

The forest vegetation consists of deciduous broad leaf trees, i. e. *Ulmus davidiana* var. *japonica*, *Pterocarya rhoifolia* and *Populous suaveolens*, in a belt along the Azusa River and coniferous forest that is located consecutively on slope of the mountains, i. e. growing *Tuga diversifolia* and *Abies veitchii*. Fruiting bodies of myxomycetes were surveyed on the decaying logs of those different tree types in the forest floor throughout 2011 – 2013 field seasons.

Myxomycete species appeared on both types of deciduous broad leaf logs and coniferous logs, associating with the decay state in normal distribution with moderately peaked stage. Ninety-two species (with varieties treated as species) from 1579 samples in total were recorded and reached 90% of estimated species richness. Species richness was 61 species in summer and 63 species in autumn. The broadleaf logs yielded 60 species and the coniferous logs were 68 species. β-diversity of myxomycetes was higher between summer and autumn, i. e. value of 0.640 in the SØrensen dissimilarity indices, than between the different tree types of broadleaf logs and coniferous logs, i. e. that of 0.540. Seasonality of species occurrence was recognized on 17 species in summer and 15 species in autumn. Preference for wood types was emerged out on 14 species for deciduous wood and 10 species for coniferous wood. The myxomycete assemblages were ordinated using non-metric multi-dimensional scaling (NMDS) in order to compare both woods types in different season. The first NMDS-axis expressed seasonal distribution and the second axis corresponded to the difference of woods types. The assemblages indicated no remarkable difference among woods types in summer time but separated clearly into distinctive groups of deciduous woods and coniferous woods in autumn. Species of Trichiales appeared dominantly on deciduous wood and species of Stemonitales emerged abundantly in coniferous woods in autumn.

Distribution pattern on decayed wood state was different between summer and autumn. Most species separately occurred on hard wood and (or) softer wood in summer, i. e. *Cribraria species* occurred on softer wood, contrary to *Physarum*. On the other hand intermediated decaying wood intensively yielded many species in autumn; especially abundant was the occurrence of *Trichia decipiens* and *Lamproderma columbinum*.

The present study demonstrated that several species seasonally use particular decayed states of wood and have substrate specificity for wood types. Thus the primitive forest in a natural conservation area of Kamikoch in central Japan furnishes myxomycete species diversity that depends on the different vegetation types in a forest.

Key words: Coniferous wood; deciduous broadleaf wood; natural conservation; primitive forest; seasonality; substrate preference

西藏地区团毛菌目黏菌

李姝,王琦

吉林农业大学食药用菌教育部工程研究中心,吉林长春

摘要:2012 至 2013 年间,在西藏地区进行黏菌野外资源考察,通过野外采集及湿室培养共获得 142 份团毛菌目黏菌标本。经分类鉴定研究,获得团毛菌目黏菌 28 种,分属于 2 科 6 属,其中 6 种为西藏新纪录种,并增加了部分团毛菌目黏菌新的分布区域。

关键词:黏菌;西藏;团毛菌目

Myxomycetes of Trichiales in Tibet

Shu Li, Qi Wang

Engineering Research Center of Chinese Ministry of Education for Edible and Medicinal Fungi, Jilin Agricultural University, Changchun, China

Abstract: Twenty eight species belonged to 2 families and 6 genera from 142 specimens of Trichiales were reported by field collection and moist chamber culture from 2012 to 2013. From them, 6 species are new records to Tibet, and some are new distribution areas.

Key words: Myxomycetes; Tibet; Trichiales

一、概述

西藏自治区位于我国西南部(东经 78°24′~99°06′,北纬 26°52′~36°32′),平均海拔 4000 米以上,地势形态复杂,在西藏自治区的东南部分布着丰富的森林资源,降水充沛(中国科学院青藏高原综合科学考察队 1988),为黏菌的生长

提供了适宜的环境条件。

团毛菌目黏菌种类繁多,子实体形态多样,多着生在腐木、枯枝上,子实体既有有柄或无柄孢囊、联囊体,也有假复囊体或复囊体;子实体无囊轴;孢丝线状,中空或实,光滑或有纹饰,末端游离或连着;孢子成堆时色浅、鲜艳,透射光下浅色至黄色、橙色或红色,孢子纹饰有刺、疣、细网纹、脊等(李玉等 2007)。

目前,针对西藏地区的黏菌资源调查还不够全面,已报道和记录的西藏黏菌共 75 种 1 变种(臧穆等 1966;图力古尔和李玉 2001;Chen et al. 2010),其中包括团毛菌目黏菌 22 种。本研究通过野外采集与湿室培养共获得西藏地区团毛菌目黏菌 28 种,西藏新纪录种 6 种,为西藏地区黏菌多样性研究增加基础数据。

二、材料方法

(一)湿室培养

培养基物为枯树枝、树皮、朽木段、落叶等(2012 – 2013 年,采集于西藏部分地区)。将不同地区采集的基物分别放入铺有滤纸的塑料盘中,喷洒无菌水,加盖塑料布保持环境潮湿,再覆上报纸避免阳光直射,室温条件培养。观察并记录黏菌的个体发育情况,待子实体成熟后,将其从基物上取下,干燥,鉴定(朱鹤等 2013)。

(二)分类鉴定

采用传统分类学方法,参照《中国真菌志—黏菌卷》(李玉等 2007)进行形态鉴定。

三、结果

经 2012 – 2013 年对西藏地区黏菌进行野外采集及湿室培养,共获得团毛菌目黏菌 142 份。其中果形团网菌 *Arcyria pomifromis* (Leers) Rostaf.、细柄半网菌 *Hemitrichia calyculata* (Speg.) Farr、棒形半网菌 *Hemitrichia clavaca* (Pers.) Rostaf.、紫褐变毛菌 *Metatrichia folriformis* (Schwein.) Nann. – Bremek.、栗褐团毛菌 *Trichia botrytis* (J. F. Gmel.) Pers.、朦纹团毛菌 *Trichia contorta* (Ditmar) Rostaf. 为西藏新纪录种(*)。结合已有报道(图力古尔和李玉 2001;Chen et al. 2010),初步记录西藏团毛菌目黏菌名录如下(△为新分布地点)。

1. *Arcyria affinis* Rostaf., Mon., 276, 1875.

生境:腐木

分布:波密(Chen et al. 2010)

2. *Arcyria brunnea* Nann. – Bremek. & Y. Yamam., Proc. K. Ned. Akad. Wet. Ser. C, 89: 219, 1986.

生境:腐木

分布:波密(Chen et al. 2010)

3. 灰团网菌 *Arcyria cineara* (Bull.) Pers., Syn. Fung., 184, 1801.

生境:腐木,苔藓,枯枝,树皮

标本:T32091,T32100,T32823,T32675,T32120

分布:米林(Chen et al. 2010),林芝,波密,察隅,墨脱$^\triangle$,昌都$^\triangle$

4. 暗红团网菌 *Arcyria denudate* (L.) Wettst., Verh. Zool. – Bot. Ges. Wien., 35: 535, 1886

生境:腐木,树皮,苔藓

标本:T32524,T32090,T32092,T32676

分布:波密,米林,林芝$^\triangle$,墨脱$^\triangle$

5. 锈色团网菌 *Arcyria ferruginea* Sauter, Flora, 24: 316, 1841

生境:腐木

分布:米林(Chen et al. 2010)

6. 粉红团网菌 *Arcyria incarnate* (Pers. ex J. F. Gmel.) Pers., Obs. Myc. 1: 58.1796.

生境:腐木,苔藓,枯枝,落叶

标本:T32093,T32101

分布:米林(Chen et al. 2010),林芝,波密$^\triangle$

7. 鲜红团网菌 *Arcyria insignis* Kalchbr. & Cooke, *in* Kalchbr., Grevillea, 10: 143, 1882.

生境:腐木

分布:波密(Chen et al. 2010)

8. 大垂网菌 *Arcyria magna* Rex, Proc. Acad. Phila. 45: 365, 1893.

生境:死树皮

分布:波密(Chen et al. 2010)

9. 大团网菌 *Arcyria major* (G. Lister) Ing, Trans. Broome Mycol. Soc. 50: 556, 1967.

生境:腐木

分布:波密(Chen et al. 2010)

10. 黄垂网菌 *Arcyria obvelata* (Oeder) Onsberg, Mycologia 70:1286,1978.

生境:腐木

标本:T32094

分布:米林(Chen et al. 2010),林芝△

11. 暗红垂网菌 *Arcyria oerstedtii* Rostaf., Mon. 278, 1875.

生境:死树皮

分布:波密(Chen et al. 2010)

12. *果形团网菌 *Arcyria pomifromis* (Leers) Rostaf., Mon. 271, 1875.

生境:枯枝,腐木,死树皮

标本:T32095,T32525,T32825

分布:林芝,波密,然乌

13. 朦纹团网菌 *Arcyria stipata* (Schwein.) Lister, Mycet. 189, 1894.

生境:腐木

标本:T32121

分布:林芝(图力古尔等 2001),昌都△

14. 异色团网菌 *Arcyria versicolor* Phill., Grevillea 5:115.1877.

生境:腐木

分布:波密,隆子(Chen et al. 2010)

15. *Calomyxa metallica* (Berk.) Nieuwl., Am. Midl. Nat., 4:335,1916.

生境:腐木

分布:米林(Chen et al. 2010)

16. *细柄半网菌 *Hemitrichia calyculata* (Speg.) Farr, Mycologia 66:887,1974.

生境:腐木,树皮,苔藓

标本:T32096,T32102,T32122

分布:林芝,波密,昌都

17. *棒形半网菌 *Hemitrichia clavaca* (Pers.) Rostaf. *in* Fuckel, Jahrb. Nass. Ver. Nat. 27-28:75,1873.

生境:腐木

标本:T32103,T32123

分布:波密,昌都

18. 蛇形半网菌 *Hemitrichia serpula* (Scop.) Rostaf. *in* Lister, Mycet. 179, 1894.

生境:腐木,苔藓

标本：T32104

分布：米林，察隅（Chen et al. 2010），波密△

19. *紫褐变毛菌 *Metatrichia folriformis* (Schwein.) Nann. – Bremek., Proc. K. Ned. Akad. Wet. C. 88：127，1985.

生境：腐木

标本：T32501，T32526

分布：林芝，波密

20. 暗红变毛菌 *Metatrichia vesparium* (Batsch) Nann. – Bremek., Proc. K. Ned. Akad. Wet. C. 69：348，1966.

生境：腐木

标本：T32502

分布：林芝

21. 小盖碗菌 *Perichaena minor* (G. Lister) Hagelst., Mycologia 35：130，1943.

生境：腐木，枯枝

分布：波密，米林，墨脱（Chen et al. 2010）

22. *栗褐团毛菌 *Trichia botrytis* (J. F. Gmel.) Pers., Neues Mag. Bot. 1：89，1794.

生境：腐木

标本：T32097，T32105

分布：林芝，波密

23. *朦纹团毛菌 *Trichia contorta* (Ditmar) Rostaf., Mon. 259，1875.

生境：腐木

标本：T32124

分布：昌都

24. *Trichia crenulata* (C. Meyl.) Bull., Soc. Vaud. Sci. Nat., 57：47，1929.

生境：死树皮

分布：林芝（Chen et al. 2010）

25. 长尖团毛菌 *Trichia decipiens* (Pers.) T. Macbr., N. Am. Slime-Moulds 218，1899.

生境：腐木

标本：T32098，T32106

分布：米林（Chen et al. 2010），亚东（图力古尔和李玉 2001），林芝△，波密△

26. 网孢团毛菌 *Trichia favoginea* (Batsch) Pers., Neues Mag. Bot. 1: 90, 1794.

生境:腐木

标本:T32099

分布:波密,林芝△

27. 鲜黄团毛菌 *Trichia lutescens* (Lister) Lister, J. Bot. 35: 216, 1897.

生境:死树皮,腐木

标本:T32125,T32824

分布:波密(Chen et al. 2010),昌都△,察隅△

28. 环壁团毛菌 *Trichia varia* (Pers.) Pers., Neues Mag. Bot. 1: 90, 1794.

生境:腐木

分布:林芝(Chen et al. 2010)

四、讨论

西藏地区生态类型多样,植被种类繁多,黏菌物种多样性随海拔增高而呈下降的趋势(Novozhilov et al. 2001)。然乌、八宿、左贡等地区多为高原草原及高山草甸生态类型,海拔 3800~5000 米,气候干燥,降雨较少,黏菌分布较少;林芝、波密、察隅等地多为原始针阔叶混交森林,海拔 2000~4200 米,黏菌多生长在气候潮湿、腐殖质层较厚的森林中。Stephenson 认为,黏菌的物种组成与森林类型组成存在着极高的一致性(Stephenson 1988),团毛菌目黏菌易着生于阔叶林及针叶林中腐木、枯枝、树皮上,主要着生树种为杨树、高山栎、落叶松、冷杉、云杉等,而在高山杜鹃灌木林中,则未发现该目黏菌,可能与该生态类型中的低温、强紫外线照射有关。黏菌的种群分布与其生活基物的种类相关(Schnittler 2001)。该目黏菌在我国东北部长白山区主要基质为落叶松、杨树、柳树,相较之下,西藏地区团毛菌目黏菌的着生基质有很大改变。

另外,通过对枯木、落叶进行湿室培养获得了多种黏菌,补充野外采集数据。以上调查结果增加了该地区黏菌的物种多样性及其地理分布数据,对于西藏地区黏菌着生基物及在高海拔地区的分布的选择性还有待进一步研究。

五、致谢

本文由国家自然科学基金项目(31370065),国家科技支撑项目课题(2012BAC01B04)支持。

参考文献

Chen, S, L., Yan, S. Z., Li, Y. 2010. An annotated checklist of Myxomycetes from Tibet, China. Mycosystema, 29(6): 845-851.

Novozhilov, Y. K., Schnittler, M., Rollins, A. W., et al. 2001. Myxomycetes from different forest types in Puerto Rico. Mycotaxon, 77: 285-299.

Schnittler, M. 2001. Ecology of myxomycetes of a winter-cold desert in western Kazakhstan. Mycologia, 93(4): 653-669.

Stephenson, S. L. 1988. Distribution and ecology of Myxomycetes in temperate forest I. patterns of occurrence in the upland forests of southwestern Virginia. Canadian Journal of Botany, 66: 2187-2207.

李玉, 李惠中, 王琦, 等. 2007. 中国真菌志-黏菌卷I. 北京: 科学出版社.

图力古尔, 李玉. 2001. 西藏真菌增补. 植物研究, 21(2): 91-194.

臧穆, 李滨, 郗建勋. 1996. 横断山区真菌. 北京: 科学出版社.

中国科学院青藏高原综合科学考察队. 1988. 西藏植被. 北京: 科学出版社.

朱鹤, 李姝, 宋晓霞, 等. 2013. 内蒙古樟子松林黏菌资源报道. 东北林业大学学报, 41(1): 124-128.

First Report of Sporangia of Two Myxomycetes (*Stemonaria longa*, *Stemonitis splendens*) Collected from Shiitake Cultivation

Bo Zhang, Shicui Jiang, Yu Li

Engineering Research Center of Chinese Ministry of Education for Edible and Medicinal Fungi, Jilin Agricultural University, Changchun, China

Abstract: Specimens of Myxomycetes collected from *Lentinulla edodes* cultivation in Henan, China, were examined. Two species of myxomycetes, *Stemonaria longa* and *Stemonitis splendens*, were identified. In the two species, occurrence of *S. longa* has become the most serious problem. This is the first report that Myxomycetes cause mushroom cultivation disease.

Key words: Amoebozoa; morphogenesis; SEM; taxonomy

1 Introduction

The shiitake mushroom, *Lentinula edodes*, is widely cultivated and managed throughout China. Henan Province is one of the highest yield of Shiitake in China.

Myxomycetes are a small, relatively homogeneous group of eukaryotic organisms and common inhabitants of decaying plant material found throughout the world. They are particularly abundant in forested regions where decaying logs, stumps, and dead leaves furnish a plentiful supply of potential substrates

In the present study, two species of myxomycetes newly found in shiitake cultivation in Henan Province, China are described. The specimens were deposited in the Herbarium of Mycological Institute of Jilin Agricultural University (HMJAU), Changchun, China. The specimens were collected from shiitake cultivation in Henan Province, China in October, 2013. Species of myxomycetes were identified based on the morphological characteristics.

2 Materials & methods

2.1 Collecting

Specimens of two species used in this study were collected from Shiitake cultivation in Sanmenxia, Henan Province, China in October 2013. They were preserved in the HMJAU. Collecting sites were 34 – 35° C at temperature and 111 – 113° W at location. Collections were glued into herbarium boxes and dried in situ.

Fig. 1 A, Artificial-log-cultivation of Shiitake were discarded because of contamination from myxomycetes (*Stemonitis splendens* or *Stemonaria longa*); B, Plasmodium on the ground in the cultivated shed; C, Transverse artificial-log-cultivation of Shiitake that had been contaminated by myxomycetes; D, Plasmodium of *Stemonaria longa* aggregated on the artificial-log-cultivation of Shiitake. (bar B = 1 cm; C = 10 cm; D = 5 cm)

2.2 Morphological studies

The fruiting bodies and microscopic structures were examined by light and scanning electron microscopes (Martin & Alexopoulos 1969). Permanent slides were mounted in Hoyer's reagent (Martin & Alexopoulos 1969). We prepared them according to Robbrecht (1974) by spreading capillitia in a drop of 94% alcohol, determining colour after one minute, and then mounting in Hoyer's reagent. Colour terms are given according to *Flora of British Fungi* (Royal Botanic Garden Edinburgh 1969).

We observed more than ten sporocarps under a stereomicroscope (20 ×) and more than 20 spores under an optical microscope (100 ×). The sporophores, capillitia, and spores were measured using a Nikon DM1000 microscope and photographed with a Canon G15 camera. For ultrastructural observation, the sporophores were attached to the holder, coated with gold using a Hitachi E – 1010 sputter, and examined with a Hitachi S – 4800 scanning electron microscope at 10 kV at Changchun Institute of Applied Chemistry, Chinese Academy of Sciences. The specimens were deposited in the HMJAU.

3 Results

3.1 Description

3.1.1 *Stemonaria longa*(Peck) Nann. – Bremek., Y. Yamam. & R. Sharma, Nederlandse Myxomyceten (Amsterdam): 505 (1983)

Morphology of dry specimen (*Stemonaria longa*) collected from artificial-log-cultivated shiitake (Fig. 2A and B) is described as follows: Sporocarps stalked, gregarious or crowded, flexous, black, long-cylindrical, 5 – 20 mm long; hypothallus silvery shinning. Stalk black (Fig. 2C – D), hollow, indistinctly longitudinally striate or homogeneous; columella tapered, almost reaching the top of the sporocarps. Capillitium open (Fig. 2 C – D), dark brown, lax, the threads rather thin, forked, free ends long, with prominently spinose tips; spore-mass black. Spores verrucose-reticulate (Fig. 2E – F), 7.5 – 9 μm in diameter; plasmodium white.

3.1.2 *Stemonitis splendens* Rostaf., Sluzowce monogr. 195 (1874)

Morphology of dry specimen (*Stemonitis splendens*) collected from artificial-log-cultivated shiitake (Fig. 3A) is described as follows: Sporocarps stalked, tufted, 10 – 17 mm long, dark brown to black; hypothallus silvery shinning; Stalk black. Columella black (Fig. 3B), tapered, almost reaching the sporocarps apex, flexous towards tip; Capillitium dark brown, the primary branches expanding at the junctions within the net and arising to the columella (Fig. 3B), the surface net with large (Fig. 3C), rounded meshes, 20 – 60 (– 110) μm in diameter. Spore-mass dark brown. Spores lilac-brown, 7 – 9 μm in diameter (Fig. 3D), warted; plasmodium white.

Sporangia of Two Myxomycetes (*Stemonaria longa* & *Stemonitis splendens*)   111

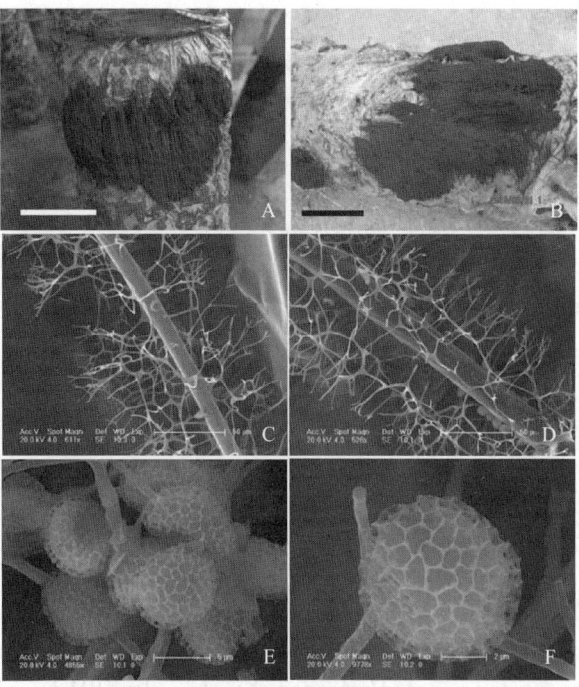

Fig. 2 *Stemonaria longa* in SEM.

A and B, Fruiting bodies on the artificial-log-cultivation of shiitake; C – D, Part of columella and capillitium; E, Many spores and part of branched capillitium; F, One spore decorated reticulate and part of branched capillitium. (bar, A, B = 10 cm)

Fig. 3 *Stemonitis splendens* in SEM.

A, Fruiting bodies on the artificial-log-cultivation of shiitake; B, Part of columella and surface net; C, Part of surface net; D, Some spores decorated warted. (bar, A = 8 cm)

References

Martin, G. M., Alexopoulos, C. J. 1969. The myxomycetes. Iowa City University of Iowa Press.

Robbrecht E. 1974. The genus *Arcyria* Wiggers in Belgium. Bulletin Natural Plantentuin Belgium, 44: 303 – 353.

Royal Botanic Garden, Edinburgh. 1969. Flora of British Fungi: colour identification chart. Edinburgh, H. M. Stationery Off.

Bo Zhang

Bo Zhang is an experimentalist at Jilin Agricultural University.

In 2004, Bo Zhang graduated from Jilin Agricultural University, with a bachelor's degree in agronomy, a master's degree in 2007 and presently is a Ph. D student in mycology under the guidance of Professor Yu Li. In 2012, she began to study taxonomy and phylogeny of true slime mold, and has published 5 relevant papers.

第四部分
Part IV

分类学
Taxonomy and Systematics

Myxomycetes of Mahe Island in the Seychelles

Tetyana Kryvomaz[1], Alain Michaud[2], Steven Stephenson[3]

1. Kyiv National Construction and Architecture University, 31, Povitroflotskyi Ave., Kyiv 03680, Ukraine;
2. 93 route de la Croizette, F - 38360 France;
3. University of Arkansas, Fayetteville, Arkansas 72701, USA

Abstract: The first study of the myxomycete biota of Mahe Island in the Seychelles was carried out in October 2011. Twenty-eight species of myxomycetes were identified from field collections, and 21 species were recovered from moist chambers cultures prepared with the bark of living lianas. In total, 43 species were recorded.

The first study of the myxomycete biota was carried out during six days (9 - 14 October 2011) in nine different localities on Mahe Island, the largest island of the Seychelles archipelago, located in the western Indian Ocean. From 100 specimens collected in the field, 28 species of myxomycetes were collected and identified by Alain Michaud and Tetyana Kryvomaz. In addition, samples of the bark from living lianas were collected for preparation of moist chamber cultures. The authors recorded 21 species from 97 specimens of myxomycetes recovered from these moist chambers cultures. In total, 43 species were recorded, and seven of these (*Arcyria cinerea*, *Physarum compressum*, *Ph. crateriforme*, *Ph. lakhanpalii*, *Ph. melleum*, *Fuligo cinereum*, and *Didymium nigripes*) were recorded as both field collections and collections from moist chamber cultures. The highest frequency of occurrence in the field was noted for *Physarum lakhanpalii* (10 specimens), whereas the most common species in moist chambers were *Physarum compressum* (21), *Collaria arcyrionema* and *Perichaena dictyonema* (both represented by 14 specimens). The vegetation of the Seychelles archipelago is marked by nearly 2000 species of tropical plants and some myxomycetes were found on dead leaves. These were *Diachea bulbillosa*, *D. leucopodia*, *Diderma effusum*, *Physarum bogoriense*, *Ph. compressum*, *Ph.*

hongkongense, *Ph. melleum*, and *Ph. mutabile*. The most common substrata for *Diderma effusum*, *D. chondrioderma*, *Perichaena corticalis* and *Physarum lakhanpalii* were the wood and bark of living coconut palm trees, and *Perichaena quadrata* was found on decayed palm wood. *Arcyria cinerea*, *A. insignis*, *Cribraria intricata*, *Lycogala epidendrum*, *Physarum bogoriense*, and *Ph. crateriforme* were collected from the dead wood of various kinds of trees, whereas *Diderma chondrioderma* was associated with mosses.

Key words: Myxomycetes; tropics; island biogeography

Quantitative Taxonomy?
—An Approach for Automated Analysis of Spore Ornamentation from SEM Images

Martin Schnittler[1], **Anna Ronikier**[3], **Paulina Janik**[3], **Yuri K. Novozhilov**[2]

1. Institute of Botany and Landscape Ecology, Ernst Moritz Arndt University Greifswald, Grimmer Str. 88, D - 17487 Greifswald, Germany;
2. Institute of Botany, Polish Academy of Sciences, Lubicz 46, 31 - 512 Krakow, Poland;
3. V. L. Komarov Botanical Institute of the Russian Academy of Sciences, Prof. Popov St. 2, 197376 St. Petersburg, Russia

Abstract: Within the last decade an increasing number of myxomycete species was described as new to science, often based on subtle morphological details derived from scanning electron microscope (SEM) micrographs. Indeed, first case studies let expect that most of these descriptions will turn out to be justified by molecular results. However, combined morphological and molecular studies showed that eye catching characters like presence or absence of a stalk (*Lamproderma/Diacheopsis*) or a capillitium (*Alwisia bombarda/A. morula*) may be misleading, whereas spore ornamentation seems to be rather more reliable. For comparisons of morphological and molecular data, a quantitative analysis of spore ornamentation would be desirable. Besides obvious figures like size and height of ornaments, we propose the following parameters to be used in quantitative investigations: (1) *coverage*: proportion of surface covered by ornamentations, (2) *density*: number of ornaments per surface area (density), (3) a) *size* and b) *shape*: mean values for all analyzed ornaments, (4) *evenness*: how evenly ornaments are spread over the spore surface, (5) degree of *reticulation*: to which degrees single ornaments are connected to a reticulum.

Image processing algorithms implemented in the freeware software ImageJ (http://imagej.nih.gov/ij/, NIH, Bethesda, Maryland) can be applied to derive estimations of these parameters.

On the example of the genus *Meriderma* we show that contrast-rich SEM images of spores, especially if prepared with the critical point drying method, are well suited for an automatized approach. This allows to separate ornaments and remaining spore surface by differences in pixel saturation. The resulting binary images can be analyzed with three basic algorithms: measuring area/shape of ornaments; counting contiguous and artificially separated ornaments (like warts connected to chains) or the meshes in-between them, and Dirichlet tessellation to estimate how evenly these ornaments are distributed. Finally, we present results of an analysis for the genus *Meriderma*.

Key words: Spores; critical point drying method; multivariate analysis

Paulina Janik

Polish Academy of Sciences.

Education

2013 – now, PhD student, Polish Academy of Sciences, International Doctoral Studies in Natural Sciences at the Polish Academy of Sciences in Kraków, Faculty of Biology.

Participation in research projects

3 projects.

Lists of publications

1. Publications in science journals: 1 paper
2. Abstract from conferences: 9 papers

Taxonomy, Phylogeny, and Morphological Evolution of the *Polysphondylium pallidum-P. album* Complex (Dictyosteliomycetes)

Shinichi Kawakami

Yamagata Prefectural Museum, Japan

Abstract: The taxonomy of Dictyostelid cellular slime molds (DCSMs) is confusing because their morphology is very simple. The genus *Polysphondylium* is phylogenetically problematic since the type species, *P. violaceum* is known to be phylogenetically separated from the most common members of the genus, the *P. pallidum* and their related species. Furthermore, a phylogenetic group including *P. pallidum* and its allied species, *P. album* (the *P. pallidum-P. album* complex) defined by a phylogenetic tree based on small subunit ribosomal DNA (SSU rDNA) has been also taxonomically confused at species level. And then, *P. pallidum* and *P. album* were redefined recently.

In this study, in order to taxonomically reevaluate this complex and the genus *Polysphondylium* and to infer phylogeny and morphological evolution of this complex, morphological, mating, and molecular phylogenetic analyses were carried out.

Firstly, based on the redefinition of *P. pallidum* and *P. album*, other described species and several isolates were systematically compared, resulting in the finding of 8 new taxa (morphospecies, tentatively named *P.* sp. 1 - 8). Morphological differences were clearly shown among 12 described species and 8 new morphospecies belonging to this complex, but *P. tikaliensis* and *P. colligatum* were morphologically similar mainly in having numerous nodes and small spores. In all the new morphospecies, macrocyst formation was observed. Mating systems of *P.* sp. 1, 3 - 7 were heterothallic and had two mating types. *P.* sp. 2 had three mating types exceptionally. On the other hand, *P.* sp. 8 produced macrocysts by itself, namely, this taxon was homothallic. By interspecies mating between all the species, a small number of macrocysts were found on some combinations. However, *P. anisocaule*,

P. pseudocandidum, and *P.* sp. 5 formed a large number of macrocysts on each combination and were morphologically similar in forming the violaceum-type aggregation.

Next, the *P. pallidum-P. album* complex was examined phylogenetically at species level. The partial nucleotide sequences of D1/D2 region of large subunit ribosomal DNA (LSU rDNA) were determined for a total of 53 strains of 14 species (13 described species of *Polysphondylium* and *D. gloeosporum*) and 8 new morphospecies of this complex. Phylogenetic tree based on these sequences showed that all the species were shown to be monophyletic. In addition, *P. anisocaule*, *P. pseudocandidum*, and *P.* sp. 5 formed a monophyletic clade. D1/D2 sequences of the morphologically similar species, *P. tikaliensis* and *P. colligatum* were the same and thus two species are considered to be same species. In addition, it is possible that three species, *P. anisocaule*, *P. pseudocandidum*, and *P.* sp. 5 belong to same one. Morphological, mating, and phylogenetic analyses revealed that 9 described species and 7 new morphospecies are distinct species. The *P. pallidum-P. album* complex was morphologically different from the type species, *P. violaceum* besides phylogenetically. Therefore, I propose that this complex is transferred to a new genus, *Oxysphondylium*.

Finally, in order to construct more resolved tree and discuss about morphological evolution on the *P. pallidum-P. album* complex, phylogenetic analyses based on SSU plus LSU rDNA sequences were carried out. The phylogenetic tree revealed the presence of two major clades within the *P. pallidum-P. album* complex. The species with larger whorl index, considerably large number of nodes, and elongation of terminal segments were detected on both the clades. Therefore, these characteristics were considered to have arisen by parallel evolution. Only *D. gloeosporum* does not have whorled branches. Therefore, it is considered that the whorls secondarily diminished during the divergence of *D. gloeosporum*. As a result of the analysis of reconstruction of ancestral states, it was suggested that the common ancestor of the *P. pallidum-P. album* complex had low whorl index, small number of nodes, and unelongated terminal segments, and aggregation of the mucoroides-type.

Key words: Taxonomy; phylogeny; evolution; polysphondylium; cellular slime molds

Dictyostelids from Jilin Province, China

Pu Liu, Yu Li

Engineering Research Center of Chinese Ministry of Education for Edible and Medicinal Fungi, Jilin Agricultural University, Changchun, China

Abstract: Dictyostelid cellular slime molds (dictyostelids) are microscopic organisms that occur in the soil and leaf litter of fields and forests soils as well as being associated with animal dung. Dictyostelids feed mostly on bacteria. They are difficult to observe and collect in the field because they are microscopic. Dictyostelids are ideal organisms suitable for investigating problems in genetics, cytology and developmental biology because of their unique macroscopic characteristics and simple life cycles.

Jilin Province is situated in the middle of northeast China and located between 122° – 131° E and 41° – 46° N which belongs to the monsoon climate of medium latitudes. The primary objectives of the present study were to find more dictyostelids from Jilin Province and compare them with known species. Eleven species of dictyostelid cellular slime molds (dictyostelids) in two genera were isolated from soil samples collected from Jilin Province in China. There are four new records for China *Dictyostelium longosporum*, *D. multistipes*, *D. gracile* and *Polysphondylium tenuissimum*. *D. clavatum* and *D. brefeldianum* had been only isolated from Taiwan of China, and another three are known species *Dictyostelium giganteum*, *Dictyostelium mucoroides* and *Polysphondylium violaceum*.

Key words: Dictyostelid cellular slime molds; *Dictyostelium*; *Polysphondylium*; taxonomy

This work was supported by the National Natural Science Foundation of China (Project Nos. 31170012, 31093440, 31300016) and Science and Technology Development Programme of Jilin Province (No. 20130522172JH).

Pu Liu

Associated Professor of Engineering Research Center of Chinese Ministry of Education for Edible and Medicinal Fungi, Jilin Agricultural University, Jilin, China.

Education

2001.9—2004.7, Jilin Agricultural University, Major: Herbal Medicine B.S

2004.9—2007.7, Jilin Agricultural University, Major: Herbal Medicine M.S.

2007.9—2010.7, Jilin Agricultural University, Major: Plant pathology Ph.D.

Academic Appointments

2010.8—2013.8, Lecturer, Engineering Research Center of Chinese Ministry of Education for Edible and Medicinal Fungi, Jilin Agricultural University.

2013.9—present, Associate professor, Engineering Research Center of Chinese Ministry of Education for Edible and Medicinal Fungi, Jilin Agricultural University.

Dictydiaethalium dictyosporangium sp. Nov. from China

Bo Zhang, Yu Li

Engineering Research Center of Chinese Ministry of Education for Edible and Medicinal Fungi, Jilin Agricultural University, Changchun, China

Abstract: *Dictydiaethalium dictyosporangium* is described as a new taxon being characterized by branched pseudocapillitium and spores marked with big rigs, sometimes forming incomplete banded-reticulate (about 10 – 12 μm in diameter) based on the bark surface of the dead log. Sporophores a pseudoaethalium, at maturity simulating an aethalium, effused, depressed, spreading over 22 mm, olivaceous to gray olivaceous, pulvinate, irregular in outline, more or less circular, extending up to 2.2 cm, up to 1.1 cm thick. Hypothallus shining, membranous, abundantly develop and surrounding the pseudoaethalium. Peridium single, membranous, translucent, slender and evanescent at the base, olivaceous brown in transmitted light, smooth, persistent. Clumella absent. Capillitium absent. Pseudocapillitium filiform, flat, thick on one side, 3 – 5 μm wide, smooth except the thickened part which bears a row of warts, branched and anastomosed, running down to the base of pseudoaethalia, yellowish green to olivaceous green, pale yellow under transmitted light. Spores free, bright yellowish green in mass, pale yellow to colourless under transmitted light, 10 – 12 μm in diameter, marked with long rigs, sometimes forming incomplete banded-reticulate. Holotype collected from Henan Province in China was deposited in the Herbarium of Mycological Institute of Jilin Agricultural University, Changchun, China.

Key words: Myxomycetes; taxonomy; *Reticulariceae*

A New Record Species of *Polysphondylium* from China

Mingjun Zhao, Pu Liu, Ying An, Dan Li, Yu Li

Engineering Research Center of Chinese Ministry of Education for Edible and Medicinal Fungi, College of Agronomy, Jilin Agricultural University, Changchun, China

Abstract: The dictyostelids (cellular slime molds), first described by Brefeld, have been known for almost a century and a half. Three genera were included in Dictyosteliomycetes, namely *Dictyostelium*, *Polysphondylium* and *Actyostelium*. Up to the present, 21 species of *Polysphondylium* have been described in the world. Ten species of *Polysphondylium* have been reported from China before this study. In 2012, forest soils were collected for dictyostelids from Mao Mountain, Jiangsu Province, which situates in eastern China and belongs to subtropical climate. Spruces and oaks are the predominant vegetation types in Mao Mountain. A new record species of Polysphondylia, *Polysphondylium colligatum* Vadell & Cavender' was isolated from the soils. The taxonomic systems of Raper & Hagiwara were used. Sorocarps, sorophores, sori, spores, cell-aggregations and pseudoplasmodia were observed. This new record species is characterized by the high frequency coremiform sorocarps, pigmentation of the sorophores and unconsolidated polar granules. Sorogens rise up synchronously and become clustered, which is similar to *D. polycephalum* Raper. However, the latter species is without whorl branches when matured. It was worth noting that, the branches keep their verticillated pattern in this strain. However, in the original report, the branches often lose their regularity and become crowded. We suggested that this difference is likely caused by environment.

Key words: Dictyostelids; *Polysphondylium colligatum*; taxonomy

This work was supported by the National Natural Science Foundation of China (Project Nos. 31170012, 31093440, 31300016) and Science and Technology Development Programme of Jilin Province (No. 20130522172JH).

Revision of the North American *Lamproderma* (Myxomycetes) Collections from the Donald T. Kowalski's Herbarium

Anna Ronikier

Institute of Botany, Polish Academy of Sciences, Lubicz 46, 31 – 512 Kraków, Poland

Abstract: Donald T. Kowalski was one of the influential scientists who studied myxomycetes in the North American mountains. His monographic paper published in *Mycologia* in 1970 (Kowalski D. T. 1970) remains an important source of information on species diversity of the genus *Lamproderma* in the USA. However, after more than 40 years of taxonomic progress in myxomycetes, this work needs a reappraisal. A revision of original collections is also needed to approach the species concepts applied by Kowalski. The present study aims at the taxonomical revision of *Lamproderma* specimens collected by Kowalski and cited in his paper in order to clarify the diversity of species in the light of current knowledge on the genus. Out of 116 specimens cited by Kowalski (1970), 95 were examined under dissection microscope, light microscope and scanning electron microscope (SEM) (21 further specimens have not yet been loaned despite several requests sent to the UC Herbarium where the collections are most likely deposited). Results of the taxonomical revision revealed that: (i) the current treatment of some species, e. g. *Lamproderma scintillans* did not change with respect to that presented by Kowalski; (ii) Kowalski's interpretation of some species was clearly different from the original species concept, e. g. treatment of *L. fuscatum* (Ronikier A., Lado C., Meyer M., Wrigley de Basanta D. 2010); (iii) interpretation of most species changed because of progress in the taxonomy of the genus, e. g. *Lamproderma carestiae*; (iv) some collections are heterogenous and contain more than one species. As a result of the taxonomical revision of the available herbarium collections 21 species were recognized. They belong to five genera: *Comatricha*, *Diacheopsis*, *Enerthenema*, *Lamproderma* and *Meriderma*.

Key words: Amoebozoa; eumycetozoa; SEM; stemonitales; taxonomy

第五部分
Part V
种系发生和遗传学
Phylogeny and Genetics

Comparisons of Genomic DNA Extraction Methods in Myxomycetes

Pu Liu, Qi Wang, Yu Li

Engineering Research Center of Chinese Ministry of Education for Edible and Medicinal Fungi, Jilin Agricultural University, Changchun, China

Abstract: Molecular biological work of myxomycetes was limited by the difficulty of genomic DNA extraction in myxomycetes. In this study, a glass homogenizer was used to grind spores. The results showed that homogenizer was more effective and available than slides for extracting genomic DNA from spores of myxomycetes. Moreover, genomic DNA of myxomycetes was successfully extracted by Biospin Fungus Genomic DNA Extraction Kit. This method was more convenient, simple, and time saving than usual genomic DNA extraction method (CTAB method) used in recent work. The concentration and purification of DNA extracted by this method were high and it could be used for PCR amplification, RAPD analysis, or other following molecular biological experiments. Both by this extraction method and homogenizer to grind, the genomic DNAs of *Physarum oblonga* were extracted and PCR amplified by three pairs of primers (12S, ITS) at the first time, and genomic DNA of *Diderma radiatum* was firstly amplified by 12S primers.

Key words: Plasmodial slime molds; genomic DNA; PCR amplification; 12S; ITSS

1 Introduction

Myxomycetes is one of the unusual groups of primitive organisms. Taxonomic relationships among myxomycetes are difficult to define because they cannot be readily classified as fungi and animal kingdom (Ashworth & Jennifer 1975). Several molecular biological methods were used to assess phylogenetic relationships of myxomycetes varying from species levels to kingdom levels (Rusk et al. 1995; Martín et al. 2003; Estrada-Torres et al. 2005; Fiore-Donno et al. 2005; Kamono & Fukui 2006; Fiore-Donno et al. 2008). The sporocarps of myxomycetes are tiny and difficult to find by the naked eyes. And the myxomycetes' collections are so limited due to the absence of universal cultivation technique for myxomycetes. Cell walls of spores of myxomycetes are more difficult to break than Fungi because without adding liquid nitrogen, enzyme, or other cell wall breaking solutions in the process of grinding spores of myxomycetes. These might be the reasons that inhibited the progress of molecular biological research in myxomycetes.

The first step for molecular biological experiments is DNA extraction. CTAB (hexadecyltrimethy-lammonium bromide) method was normally used in Fungi and myxomycetes. Cell wall breaking is a critical factor for the success of DNA extraction. In myxomycetes, slide crashing method was used to extract DNA (Liu & Li 1996; Liu et al. 2001) enlightened by Lee and Taylor (1990). There also have been successful studies in which DNA was extracted by slide crashing method (Williams et al. 1990; Foster et al. 1993).

In this article, comparison of genomic DNA extraction methods in myxomycetes was discussed. Grinding spores by a homogenizer is more convenient and effective than slide crashing method. Genomic DNA of *Physarella oblonga* Berk. & Curt. and *Diderma radiatum* Morgan were both first succefully extracted by Fungus Genomic DNA extraction kit and CTAB method. The results of this study proved that the concentration and purification of DNA extracted by the kit is higher than by CTAB method. And the kit method is much easier and time saving than CTAB method. DNA of the two species of myxomycetes extracted by the kit were PCR amplified by three pairs of primers.

2 Materials and methods

2.1 Sampling

Sporocarps of *P. oblonga* and *D. radiatum* used in this study were collected from Zuojia and Lushuihe, Jilin province, northeast China, in September 2005. They were preserved in Mycological Herbarium of Jilin Agricultural University (HMJAU).

2.2 DNA extraction

Spores of each sample were ground by a homogenizer. Adding lysis buffer in the homogenizer during grinding could speed up lysis. DNA extraction from sporocarp samples was carried out by CTAB method and Biospin Fungus Genomic DNA extraction kit (Bioer Technology), following the instructions of the manufacturer but without adding RNase A. The extracted solution was heated for 1 h at 65℃. The CTAB method needed an overnight incubation in lysis buffer. The Biospin Fungus Genomic DNA extraction kit method needed only a half daytime in the whole process of DNA extraction. Finally, DNA was resuspended in elution sterile ultra pure water and could be used for PCR amplification.

2.3 PCR amplification

Genomic DNA of the two samples extracted by Biospin Fungus Genomic DNA extraction kit were PCR amplified by primers 12S (forward and reverse), ITS1 – ITS4, and ITS5 – ITS4 (White et al. 1990). Length of those amplified sequences are approximately 400 bp (primers 12S), 608 – 633 bp (primers ITS1 – ITS4), and 600 – 700 bp (primers ITS5 – ITS4). The thermocycling program was 30 s at 94℃, 30 s at 50 – 56℃, 1 min at 72℃ (repeated for 35 cycles) and 8 – 10 min at 72℃. DNA was amplified by Takara PCR Amplification Kit (Takara Biotechnology, Dalian, China). The PCR solution consisted of 0.25 μL of Takara Taq^{TM} (5 U/μL), 5.0 μL of 10 × PCR Buffer, 2.0 μL of dNTP Mixture (2.5 mmol/L each), 1 – 100 ng of template DNA, 1.0 μL of each primer (μmol/L), and sterile water that was adjusted to a final volume of 50 μL. The tubes were initially placed in ice and immediately into the PCR Life Express^{TM} (Hangzhou, China) thermocycler. PCR products were analyzed by 1.0% (w/v) agarose gel electrophoresis and were subsequently visualized by ethidium bromide staining.

3 Results

Spores of sample A were ground by slides and DNA extracted by CTAB method. Spores of samples B and C were ground by a homogenizer and DNA extracted by CTAB method. Spores of sample D were ground by slides and DNA extracted by Biospin Fungus Genomic DNA extraction kit. Spores of samples E and F were ground by a homogenizer and DNA extracted by the kit. The results showed that the concentration of DNA extracted using a homogenizer to grind was better than ground samples by slides in the same DNA extraction method (Fig. 1, A compared with B and C, D compared with E and F). Moreover, the purification of DNA extracted by the kit was higher than by CTAB method in the same grinding method (Fig. 1, A compared with D, B compared with E, and C compared with F). The same results were also found in *Diderma radiatum*. The length of DNA extraction time by Biospin Fungus Genomic DNA extraction kit is much shorter than by CTAB method. Using a homogenizer to grind spores and kit extraction method, genomic DNA of *Physarella oblonga* and *D. radiatum* were both successfully amplified by 12S primers (Fig. 2). Furthermore, genomic DNAs of *P. oblonga* were also amplified by two pairs of primers ITS1 – ITS4 and ITS5 – ITS4 successfully (Fig. 3).

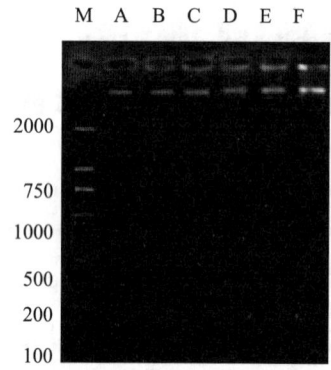

Fig. 1 Electrophoresis of DNA of *Physarella oblonga*.

The dose of each sample (A – F) was 5 μL.

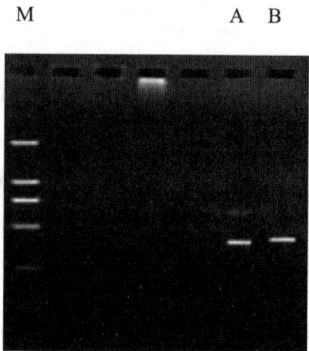

Fig. 2 Electrophoresis of PCR amplification products by primers 12S.

A, *P. oblonga*; B, *Diderma radiatum*; M, Marker.

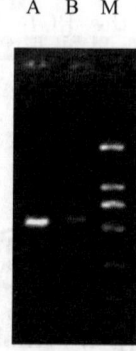

Fig. 3 Electrophoresis of *P. oblonga* PCR amplification products

A, primers ITS1 – ITS4; B, primers ITS5 – ITS4; M, Marker

4 Discussion

Fungus genomic DNA extraction kit is also suitable for Myxomycetes from this study by the good straps of DNA obtained from electrophoresis of DNA. Kit method is more convenient and time consuming, and has got large quantity of DNA than CTAB method. So, this kit could be used to extract DNA of myxomycetes. During the process of DNA extraction, the most important factor is cell wall breaking. The success of DNA extraction and quality of DNA were determined by grinding time. Using a homogenizer to grind spores could make up for the insufficiency of slide crashing method which could only grind spores. Because a whole surface sterilizing sporangium could be directly put into a homogenizer, this could prevent the inconvenience of DNA extraction of myxomycetes of less spores by slide crashing method. A homogenizer is a substitute of slides to grind spores and could enlarge the sources of DNA.

This is the first time to extract genomic DNA of *P. oblonga* and do PCR amplification. Genomic DNAs of *P. oblonga* extracted by the kit were PCR amplified by three pairs of primers. So, those three pairs of primers were all suitable for *P. oblonga*. Genomic DNA of *D. radiatum* was firstly amplified by 12S primers. From this study, *P. oblonga* and *D. radiatum* were both PCR amplified by the same pair of primers (12S). But these two nucleotide sequences were distinguishable. This result is in accordance with their morphological characteristics. *P. oblonga* has cylindric pseudocolumella and duplex capillitium composed of limy, spine-like, branched processes arising from the inner walls of the outer part of the peridium and a dense network of threads bearing a few nodes. Furthermore, *D. radiatum* has large hemispherical to subglobose columella; its capillitium is abundant, of brown threads and hardly branching. This study is also a basis for the further molecular biological research of *P. oblonga* and *D. radiatum*.

References

Ashworth, J. M., Jennifer, D. 1975. The Biology of Slime Moulds. London: Edward Arnold. 1 – 31.

Estrada-Torres, A., Gaither, T. W., Miller, D. L., et al. 2005. The myxomycete genus Schenella: morphological and DNA sequence evidence for synonymy with the gasteromycete genus Pyrenogaster. Mycologia, 97(1):139 – 149.

Fiore-Donno, A. M., Berney, C., Pawlowski, J., et al. 2005. Higher-order phylogeny of plasmodial slime molds (Myxogastria) based on elongation factor 1 – A and small subunit rRNA sequence. Journal of Eukaryotic Microbiology, 52: 201 – 210.

Fiore-Donno, A. M., Meyer, M., Baldauf, S. L, et al. 2008. Evolution of dark-spored Myxomycetes (slime-molds): molecules versus morphology. Molecular Phylogenetics and Evolution, 46:878 – 889.

Foster, L. M., Kozak, K. R., Loftus, M. G., et al. 1993. The polymerase chain reaction and its applications to filamentous fungi. Mycological Research, 97(7): 769 – 781.

Kamono, A., Fukui, M. 2006. Rapid PCR-based method for detection and differentiation of Didymiaceae and Physaraceae (myxomycetes) in environmental samples. Journal of Microbiological Methods, 67:496 – 506.

Lee, S. B., Taylor, T. W. 1990. Isolation of DNA from fungal mycelis and single spore// Innis, M. A., Gelfand, D. H, Sninsky, J. J., et al. PCR Protocols: A Guide to Methods and Applications. San Diego, California: Academic Press. 282 – 287.

Liu, S. Y., Li, Y. 1996. Studies on the method of DNA extraction from myxomycetes. Jilin Nongye Daxue Xuebao, 18(suppl.):61 – 63 (in Chinese),

Liu, S. Y., Li, Y., Bai, X. J. 2001. A new method for extracting DNA from myxomycetes. Jilin Nongye Daxue Xuebao, 23(2): 38 – 40 (in Chinese).

Martín, M. P., Lado, C., Johansen, S. 2003. Primers are designed for amplification and direct sequencing of ITS region of rDNA from Myxomycetes, Mycologia, 95(3):474 – 479.

Rusk, S. A., Spiegel, F. W., Lee, S. B., 1995. Design of polymerase chain reaction primers for amplifying nuclear ribosomal DNA from slime molds, Mycologia, 87(1):140 – 143.

White, T. J., Bruns, T., Lee, S., et al. 1990. Amplification and direct sequencing of fungal ribosomal RNA genes for phylogenetics // Innis, M. A., Gelfand, D. H., Sninsky, J. J., et al. PCR Protocols: A Guide to Methods and Applications. California, San Diego: Academic Press. 315 – 322.

Williams, J. G. K., Kubelik, A. R, Livak, K. J. et al. 1990. DNA polymorphisms amplified by arbitrary primers useful as genetic markers. Nucleic Acids Research, 18(22):6531 – 6535.

Molecular Phylogeny of Some Myxomycetes Taxa

Shuyan Liu, Fenyun Zhao, Yu Li

Engineering Research Center of Chinese Ministry of Education for Edible and Medicinal Fungi, Jilin Agricultural University, Changchun, Jilin Province, China

Abstract: Myxomycetes are one of the most diverse fungi. Their life cycle includes a multinucleate somatic phase known as a plasmodium and a reproductive phase producing sporophores with walled spores, which makes them very different from all other organisms. The molecular phylogeny of the taxa in Myxomycetes has been studied based on the sequences analysis of SSUrDNA and EF1 – α genes. But there are only limited species used in those analyses.

In order to exam the phylogenetic relationship among the groups, COI and EF1 – α gene were amplified and sequenced. The sequences were initially aligned using the Clustal X package. The alignments were manually edited using MEGA 6. Phylogenetic trees were obtained from the data by the Maximum-Parsimony method using the heuristic search option in the program PAUP* 4.0b8. This search was repeated 100 times with different random starting points, using the stepwise addition option to increase the likelihood of finding the most parsimonious tree. Transversions and transitions were treated with equal weight. All sites were treated as unordered, with gaps treated as missing data. The branch-swapping algorithm was TBR, the MULPARS option was in effect, and zero length branches were collapsed. The strength of the internal branches from the resulting trees was tested by bootstrap analysis using 1000 replications. The results showed that the four orders of myxomycetes formed three clades, Stemonitales forming a clade, Liceales and Trichales forming a clade, and Physarales forming a clade. In Physarales clade, plasmodiocarp and aethalia groups were more closed than sessile sporocarp groups, which showed that the type of fruiting bodies of slime molds may indicate some phylogenetic meaning during the morphogenesis of slime mold fruit body.

Key words:Slime mold; COI; EF1 - α; systematics

The Phylogeny of Slime Moulds (Mycetozoa): from One Gene to the Whole Genome

Cong Fu[1], Yu Li[2]

1. State Key Laboratory of Theoretical and Computational Chemistry, Institute of Theoretical Chemistry, Jilin University, Changchun, China;
2. Engineering Research Center of Ministry of Education for Edible and Medicinal Fungi, Changchun, China

Abstract: Over the past three decades, the sequencing technologies have been developed rapidly, from first-generation Sanger DNA sequencing to the current fast and cost-effective next-generation sequencing (NGS). By using these sequencing technologies, more and more complete genomes have been sequenced, which has pushed phylogenetic analysis into a new era, in particular shedding light on the complicated evolutionary history of slime molds.

The evolutionary history of a collection of organisms is usually represented by a phylogeny in the form of a tree. Phylogenies can be constructed in many ways. For example, some methods are at the sequence level, constructing phylogeny trees based on sequencing similarities; some methods are beyond the sequence level, and they use features across the whole genome. Each method has its advantages and bias; they all more or less help us to understand the history of life better and may leave some problems unsolved.

In this article, we review the recent advances in phylogeny construction. In particular the phylogenetic approaches have been applied to uncover the evolutionary history of slime molds, and discuss possible developments towards a comprehensive reconstruction of slime molds phylogeny in the future.

Key words: Phylogeny; slime moulds (Mycetozoa); next-generation sequencing

盘基网柄菌肌动蛋白保守基序的生物信息学分析

李广[1,2],刘淑艳[1],李玉[1],陈艳秋[2]

1. 吉林农业大学食药用菌教育部工程研究中心,吉林长春;
2. 延边大学农学院,吉林延吉

摘要:盘基网柄菌(*Dictyostelium discoideum*)是目前黏菌中研究最清楚的模式生物,其捕食过程与肌动蛋白的多聚化密切相关。为探讨盘基网柄菌肌动蛋白的序列特征,本研究利用 MEME SUITE 分析盘基网柄菌 32 条肌动蛋白序列,获得 motif1,motif2,motif3 3 个保守基序。同时分析了与盘基网柄菌肌动蛋白 actin17 最匹配的 21 种生物肌动蛋白,获得 Motif1,Motif2,Motif3 3 个保守基序。其中 motif1,Motif1,Motif3 为本研究新发现的保守基序,这 3 个保守基序可定位于 Profilin-actin-VASP$_{202-244}$(PDB ID:3CHW)三维结构的重要位置。以上结果表明,motif1,Motif1,Motif3 可能是盘基网柄菌肌动蛋白在进化中重要的保守基序。

关键词:黏菌;生物信息;基序分析

Bioinformatics Analysis of Conserved Motifs of Actins from *Dictyostelium discoideum*

Guang Li[1,2], Shuyan Liu[1], Yu Li[1], Yanqiu Chen[2]

1. Engineering Research Center of Chinese Ministry of Education for Edible and Medicinal Fungi, Jilin Agricultural University, Changchun, China
2. Agricultural College of Yanbian University, Yanji, Jilin Province, China

Abstract: *Dictyostelium discoideum* is widely used as a model organism in slime mold. Its predation process is closely related to actins polymerization. To investigate the characteristics of actins from *Dictyostelium discoideum*, the conserved motifs of 32 actins from *Dictyostelium discoideum* were analyzed using MEME SUITE analyzing method. The results indicated that three conserved motifs, motif1, motif2, and motif3, were found from the 32 actins sequences; three conserved motifs, Motif1, Motif2, Motif3 were found from actin17 and its 21 best hit genes of other organisms' actins. The motifs, Motif1, Motif1, and Motif3 were newly found in this study, and were located on the crystal structure of Profilin-actin-VASP$_{202-244}$. All these results suggested that motif1, Motif1 and Motif3 may play an important role in the evolution of conserved motifs.

Key words: Slime mold; bioinformation; motif analysis

一、概述

肌动蛋白在所有的真核细胞当中均有发现,在体内肌动蛋白以肌动蛋白单体(G-actin)和肌动蛋白纤维(F-actin)两种形式存在。在生物分子演化中,肌动蛋白是高度保守的蛋白质分子之一,其主要功能是构成细胞骨架中的微丝系统(Dominguez & Holmes 2011)。肌动蛋白与肌球蛋白(myosin)共同作用可以将ATP中的化学能转化成机械能,为细胞的运动提供动力,因此,肌动蛋白在细胞的各种生理活动中起着十分重要的作用。在分类学上,盘基网柄菌(*Dictyostelium discoideum* Raper)属于变形虫界(Amoebozoa)菌虫门(Mycetozoa)网柄菌纲(Dictyostelia)网柄菌科(Dictyostseliaceae),是一种生长在土壤、腐烂的蘑菇和植物材料、食草动物粪便等场所,以细菌为主要食物的微小生物。在黏菌中盘基网柄菌是研究最清楚的模式生物(Annesley and Fisher 2009)。其主要特征是孢子萌发产生单核、裸露、单倍体的变形体,变形体通过集群产生假原生质团,然后生成孢堆果,在孢堆果中形成孢子。该菌是一种具有社会性、介于简单的单细胞生物与高等的多细胞生物之间的生物类型(Raper & Rahn 1984)。

早在1941年,研究人员在脊椎动物的骨骼细胞中发现肌动蛋白(Perry 2003),在此之后,Loewy(1952)第一次从黏菌中提取到肌动蛋白。早期对盘基网柄菌肌动蛋白的研究多为生理生化方面的研究(Spudich & Lord 1974;Uyemura et al. 1978)。盘基网柄菌在吞食细菌的过程中,伪足细胞内会聚集很多肌动蛋白纤维(Condeelis et al. 1988)。已有研究证明,肌动蛋白的聚合、解聚为盘基网柄菌的运动、胞质分裂及吞噬作用等生理过程所必需(Vicker 2002)。20世纪80年代末,人们估计盘基网柄菌基因组可能含有17~20条肌动蛋白(Knecht et al.

1986），但随着盘基网柄菌基因组全测序的完成，发现盘基网柄菌基因组含有 32 条肌动蛋白（Eichinger et al. 2005），研究人员开始应用生物信息学方法分析肌动蛋白的特点，探究肌动蛋白在盘基网柄菌运动过程中发挥的作用。Josep 等（2008）比较了 41 条盘基网柄菌肌动蛋白（32 条）和肌动蛋白相关蛋白（9 条）的序列特征，其结果表明，有 17 条盘基网柄菌肌动蛋白在进化上几乎没有变化（95%的相似性），其可能是基因复制的结果。肌动蛋白也被作为分子标记基因应用于植物界、动物界和真菌界的分子系统学研究（Drouin et al. 1995；Kapoor et al. 2008；Keeling 2001），Lahr 等（2011）基于串联 SSU - rDNA 和肌动蛋白基因，全面分析重建了阿米巴系统发育。

基序（Motif）是 DNA 或氨基酸序列中短的、可能具有一定生物学功能的回环模式，通常预示蛋白质特定的结合位点，例如核酸酶和转录因子的结合位点等（D'Haeseleer 2006）。因此，通过计算机方法筛选 DNA，RNA 以及蛋白质序列功能区所包含的特征性基序是一个基础而广泛研究的课题（Frith et al. 2008）。邓荟芬等（2007）研究发现，在拟南芥保卫细胞中一些基因转录起始密码子 ATG 上游 500 bp 序列中的 AAAAG 基序和 TAAAG 基序，控制着一些基因的特异性表达。Joseph 等（2008）通过将盘基网柄菌肌动蛋白和肌动蛋白相关蛋白进行比较获得了在结构、功能以及进化上具有重要意义的 5 个保守基序。在禾谷镰刀菌（*Fusarium graminearum* Petch）和大丽轮枝菌（*Verticillium dahliae* Kleb）分泌组中，分别被预测出 157 个和 97 个含有 RxLx[EDQ]基序的潜在的病原菌-寄主互作蛋白（田李等 2011；于钦亮等 2008）。

本研究拟采用生物信息学分析方法，对盘基网柄菌 32 条肌动蛋白保守基序进行分析，为解释黏菌运动机理提供工作基础。

二、材料与方法

（一）材料

本研究所需要盘基网柄菌（*Dictyostelium discoideum*）的肌动蛋白序列和基因注释信息均由盘基网柄菌基因组数据库 DictyBase（Basu et al. 2012）（http://dictybase.org/）下载获得。

（二）方法

1）盘基网柄菌肌动蛋白序列的获得及其在染色体上的分布：所有盘基网柄菌的肌动蛋白序列通过 Perl 程序从基因组数据库获得，并将其定位在基因组染色体的相应位置上。

2）盘基网柄菌肌动蛋白基因的系统发育分析：为了探讨盘基网柄菌肌动蛋白基因间的系统发育关系，本研究采用 Clustal×1.81（Thompson et al. 1997）对盘基网柄菌肌动蛋白氨基酸序列进行比对，然后使用 MEGA5.2（Tamura et al. 2011）通过 ME 法来构建系统发育树。

3）盘基网柄菌肌动蛋白氨基酸序列的保守基序分析：为预测盘基网柄菌肌动蛋白氨基酸序列的保守基序，本研究利用 MEME SUITE（Bailey et al. 2009）分析肌动蛋白氨基酸序列，将进一步探讨保守基序在蛋白质三维结构上的相对位置。利用 PyMOL（DeLano 2002）软件渲染肌动蛋白三维结构。

三、结果与分析

（一）盘基网柄菌肌动蛋白在基因组中的分布情况

通过统计分析共获得 32 条盘基网柄菌肌动蛋白基因，分布在 5 条染色体上。其中 1 号染色体上有 5 条肌动蛋白基因、2 号染色体上有 13 条、3 号染色体上有 1 条、4 号染色体上有 1 条、5 号染色体上有 12 条。78% 的肌动蛋白基因分布在 2 号和 5 号染色体上。在 32 条盘基网柄菌肌动蛋白中，12 条是克里克链（Crick strand），20 条是沃森链（Watson strand），详细信息见表 1。

表1　盘基网柄菌肌动蛋白基因在染色体中的排列信息

Table 1　Gene neighborhoods of actin of *Dictyostelium discoideum* on chromosome

盘基网柄菌数据库序列号 DictyBase ID	蛋白 Proteins	染色体 Chromosome	坐标 Coordinates	链 Strand
DDB0220461	act23	chromosome 1	2031686 to 2032762	Watson strand
DDB0220464	act26	chromosome 1	3826453 to 3827580	Watson strand
DDB0216213	act8	chromosome 1	4602557 to 4603687	Watson strand
DDB0234013	act31	chromosome 1	2844541 to 2845608	Watson strand
DDB0234014	act32	chromosome 1	2967350 to 2968528	Crick strand
DDB0216214	act12	chromosome 2	4718776 to 4719906	Crick strand
DDB0220454	act13	chromosome 2	4495093 to 4496223	Crick strand
DDB0220455	act14	chromosome 2	4418543 to 4419673	Watson strand
DDB0185015	act15	chromosome 2	1781191 to 1782321	Watson strand
DDB0185127	act16	chromosome 2	1607842 to 1608972	Crick strand

续表

盘基网柄菌 数据库序列号 DictyBase ID	蛋白 Proteins	染色体 Chromosome	坐标 Coordinates	链 Strand
DDB0185125	act17	chromosome 2	4723516 to 4724640	Watson strand
DDB0220446	act19	chromosome 2	4053689 to 4054819	Watson strand
DDB0185124	act2	chromosome 2	4722018 to 4723148	Watson strand
DDB0220450	act20	chromosome 2	4047769 to 4048899	Crick strand
DDB0220460	act22	chromosome 2	5343575 to 5344705	Watson strand
DDB0185126	act6	chromosome 2	4055916 to 4057046	Watson strand
DDB0220456	act9	chromosome 2	4455464 to 4456594	Watson strand
DDB0220451	act21	chromosome 2	4052062 to 4053192	Crick strand
DDB0220445	act7	chromosome 3	3684122 to 3685252	Watson strand
DDB0229355	act29	chromosome 4	5035700 to 5036863	Crick strand
DDB0220444	act1	chromosome 5	2938326 to 2939456	Crick strand
DDB0220457	act10	chromosome 5	3337403 to 3338533	Watson strand
DDB0220449	act11	chromosome 5	2167534 to 2168664	Watson strand
DDB0220459	act18	chromosome 5	2886806 to 2887948	Watson strand
DDB0220458	act3	chromosome 5	2884652 to 2885782	Watson strand
DDB0234012	act33	chromosome 5	925837 to 927081	Crick strand
DDB0229354	act28	chromosome 5	2883081 to 2884142	Watson strand
DDB0220448	act4	chromosome 5	2234318 to 2235448	Crick strand
DDB0220447	act5	chromosome 5	3186992 to 3188122	Crick strand
DDB0220462	act24	chromosome 5	2876728 to 2877861	Crick strand
DDB0220463	act25	chromosome 5	2878511 to 2879668	Watson strand
DDB0229353	act27	chromosome 5	2880348 to 2881466	Watson strand

（二）盘基网柄菌肌动蛋白基因的系统进化分析

通过对盘基网柄菌肌动蛋白氨基酸序列进行多序列比对，用 MEGA5.2 通过 ME 法构建的盘基网柄菌 32 条肌动蛋白的系统发育树如图 1 所示，利用 Bootstrap test 分析来确定进化树分化节点的自展支持率，重复 1000 次。以变形虫界的溶组织内阿米巴 Entamoeba histolytica Schaudinn 的肌动蛋白为外群，其 GenBank 登录号为 BAN39822.1。

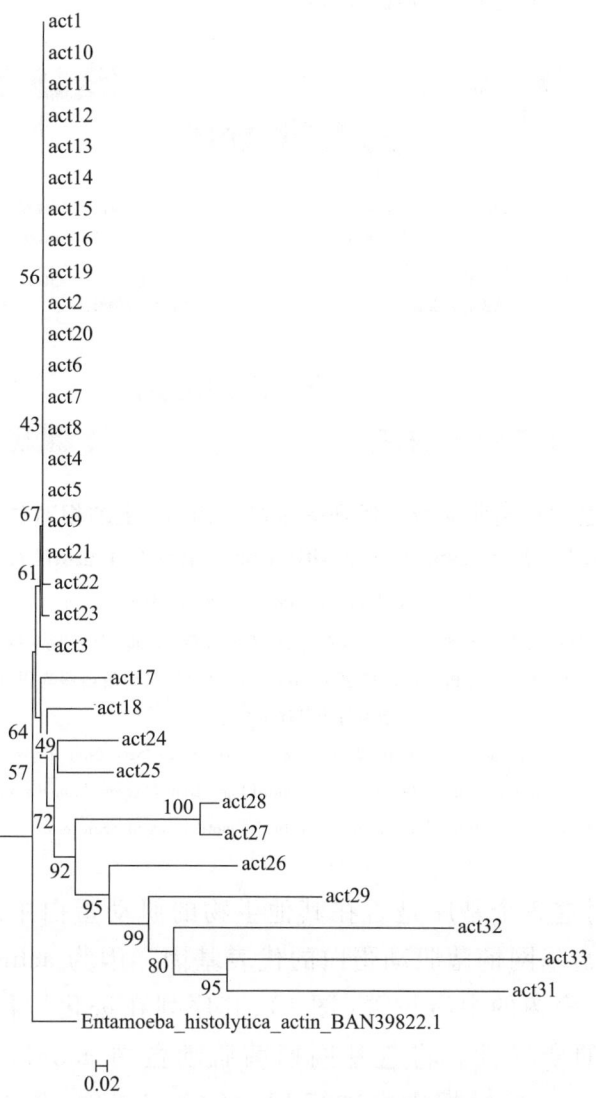

图 1 用 ME 法建立的盘基网柄菌肌动蛋白基因系统发育树
（分枝下的数字为 bootstrap 支持率）

Fig. 1 Phylogenetic tree of *Dictyostelium discoideum* actins using ME method. (Numerals below the branches show bootstrap support values with 1000 replicates.)

（三）盘基网柄菌肌动蛋白氨基酸序列的保守基序分析

利用 MEME SUITE 对盘基网柄菌 32 条肌动蛋白氨基酸序列的保守基序进行分析,得到 3 个保守基序,即位于上游的 motif1（E 值为 6.8e－924）、位于中游的 motif2（E 值为 1.8e－1280）、位于下游的 motif3（E 值为 1.7e－1267）3 个保守基序在肌动蛋白上的相对位置如图 2A。

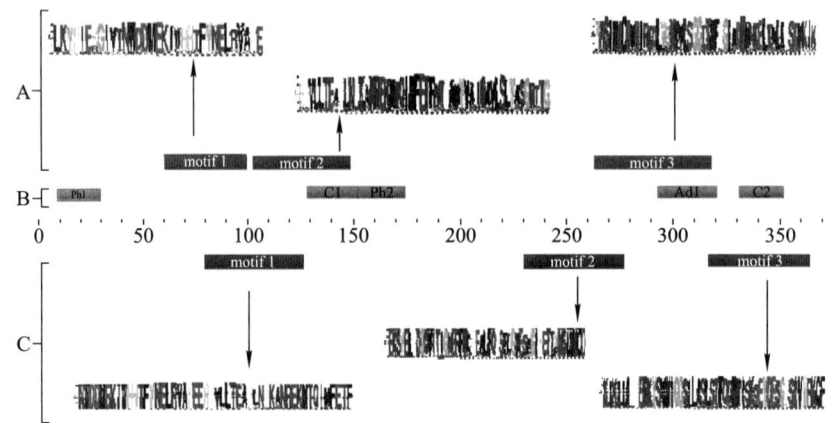

图 2 盘基网柄菌肌动蛋白保守基序在 actin 17 上的相对位置示意图
Fig. 2 The respective position on actin17 of motif from *Dictyostelium discoideum* actins.

A,分析盘基网柄菌 32 条肌动蛋白序列得到的保守基序;B,Joseph 等(2008)分析盘基网柄菌肌动蛋白与肌动蛋白相关蛋白得到的保守基序;C,分析盘基网柄菌 actin17 与其他生物肌动蛋白序列得到的保守基序(箭头所指为每个基序的标示)

A, The motif logos from 32 actins of *Dictyostelium discoideum*; B, The motif logos from *Dictyostelium discoideum* actins and actin-related proteins by Joseph et al. (2008); C, The motif logos from *Dictyostelium discoideum* actin17 with other organisms actins (The arrows point to the motif logos of each motif).

为进一步探讨这 3 个基序是否在其他生物的肌动蛋白中广泛存在,本研究选择 actin17 作为盘基网柄菌肌动蛋白的代表基因。因为 actin17 处于盘基网柄菌肌动蛋白系统发育树的中游位置(图 1),其序列在进化过程中既有一定的保守性又存在一定的变异性。将盘基网柄菌肌动蛋白 actin17 在 NCBI(http://www.ncbi.nlm.nih.gov/)数据库中进行 Blast(Altschul et al. 1990)分析,筛选出最佳匹配基因 21 个,分属于动物界、变形虫界、植物界、真菌界、细菌界,这些基因的详细信息见表 2。

同样,利用 MEME SUITE 分析盘基网柄菌 actin17 与其他生物肌动蛋白保守基序,结果如图 2C 所示,发现 3 个保守基序(本文以 Motif,motif 区分两次分析获得的基序),分别是位于上游的 Motif1(E 值为 $1.2e-1097$)、位于下游的 Motif2(E 值为 $4.0e-1009$)、位于中游的 Motif3(E 值为 $7.5e-974$),并发现图 2A 中的 motif1,motif2 与图 2C 中的 Motif1 存在重合;motif3 分别与 Motif2 存在重合。将上述 6 个保守基序与 Joseph 等(2008)分析盘基网柄菌肌动蛋白与肌动蛋白相关蛋白得到的 5 个保守基序进行比较,结果表明,motif1,Motif1,Motif3 与已报道的基序明显不同,为本研究新得到的保守基序,各基序在 acin17 上的相对位置如图 2 所示。

表 2 与 actin17 具有最佳匹配的其他生物肌动蛋白基因信息
Table 2 The information of the best hit actins gene with actin17 from other organisms

界 Kingdom	物种 Species name	GenBank 登录号 GenBank accession No.	基因产物 Gene Product
动物界 Animal	斑马鱼 Danio rerio	NP_571106.1	actin, cytoplasmic 1
	意大利蜜蜂 Apis mellifera	XP_625015.1	actin, clone 403 – like
	家牛 Bos taurus	NP_776404.2	actin, cytoplasmic 1
	果蝇 Drosophila melanogaster	NP_511052.1	actin 5C, isoform B
	原鸡 Gallus gallus	NP_990849.1	actin, cytoplasmic 1
	人类 Homo sapiens	NP_001092.1	actin, cytoplasmic 1
	小鼠 Mus musculus	NP_031419.1	actin, cytoplasmic 1
	黑猩猩 Pan troglodytes	NP_001009945.1	actin, cytoplasmic 1
	大鼠 Rattus norvecus	NP_112406.1	actin, cytoplasmic 1
变形虫界 Amoebozoa	紫网柄菌 Dictyostelium purpureum	XP_003289998.1	hypothetical protein
	轮柄菌 Polysphondylium pallidum	EFA80799.1	actin
	Dictyostelium fasciculatum	EGG14129.1	actin
	溶组织内阿米巴 Entamoeba histolytica	BAN39822.1	actin
植物界 Plantae	拟南芥 Arabidopsis thaliana	NP_187818.1	actin
	水稻 Oryza sativa	NP_001051086.1	hypothetical protein
真菌界 Fungi	黑粉菌 Ustilago maydis	XP_762364.1	actin
	灰绿犁头霉 Absidia glauca	AAA32619.1	actin – 2
	粗糙脉孢菌 Neurospora crassa	AAC78496.1	actin
	禾本科布氏白粉菌 Blumeria graminis	CCU76638.1	actin
细菌界 Bacteria	草螺菌 Herbaspirillum seropedicae	WP_017454560.1	hypothetical protein
	铜绿微囊蓝细菌 Microcystis aeruginosa	WP_002745786.1	actin, cytoplasmic 2

为进一步研究 motif1,Motif1,Motif3 3 个保守基序(图 2)在蛋白质三维结构上的相对位置,将 actin17 氨基酸序列在 NCBI 蛋白质数据库 Protein Data Bank (PDB)中进行 Blast 分析,筛选出与该序列一致性(identity)最高的蛋白结构是 Profilin-actin-VASP$_{202-244}$(PDB ID:3CHW)(Baek et al. 2008)。用 PyMOL 软件渲染后得到其三维结构图(图 3A)。将保守基序 motif1,Motif1,Motif3 在这个三维结构上标记相对位置后如图 3B – D 所示。由图 3B – D 可以看出,基序

motif1、Motif1、Motif3 都位于 α-螺旋和 β-折叠上。这表明，motif1、Motif1、Motif3 可能是盘基网柄菌肌动蛋白在进化中重要的保守基序。

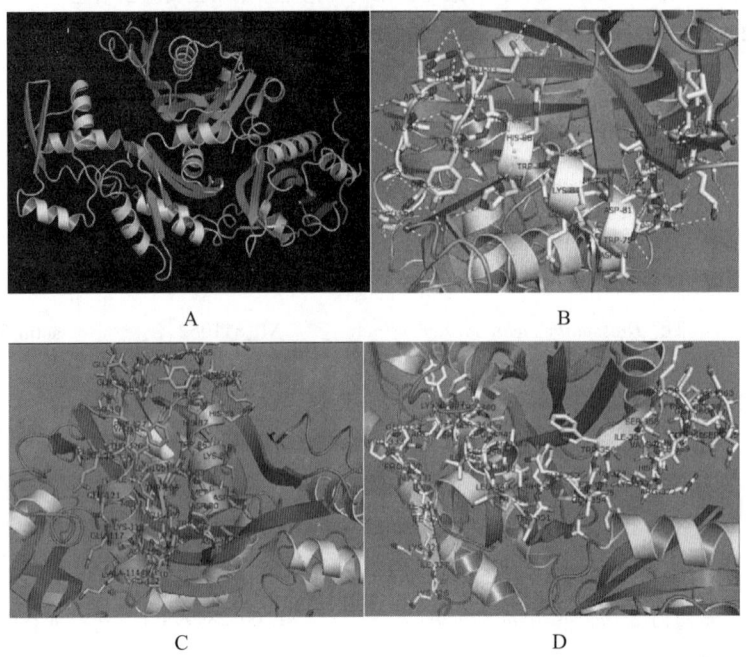

图 3　盘基网柄菌 3 个保守基序在 Profilin-actin-VASP$_{202-244}$ 蛋白复合体晶体结构上的定位

Fig. 3　The location of three motif3 of *Dictyostelium discoideum* actins on the crystal structure of Profilin-actin-VASP$_{202-244}$.

A，Profilin-actin-VASP$_{202-244}$ 蛋白复合体的晶体结构；B – D，保守基序 motif1、Motif1、Motif3 在 Profilin-actin-VASP$_{202-244}$ 蛋白复合体晶体结构上的定位

A, The crystal structure of Profilin-actin-VASP$_{202-244}$; B – D, The position of conserved motif, motif1, Motif1 and Motif3 on the crystal structure of Profilin-actin-VASP$_{202-244}$ respectively.

四、结论与讨论

基因的调控分析中，序列的基序越来越重要（D'Haeseleer 2006），在真核生物中的 C2H2 锌指域是广泛保守的，其中串联的 CWCH2 序列基序可以作为内锌指（inter-zinc finger）相互作用的标志（Hatayama & Aruga 2010）。两个高度保守的 NPA 基序作为水通道蛋白（aquaporin, AQP）的重要结构域，其对水的选择性渗透起着关键作用（Guan et al. 2010）。卵菌无毒基因中在信号肽的效应因子的 N 端存在 RXLR 和 dEER 两个基序（Birch et al. 2006；Rehmany et al. 2005），其中 RXLR 基序可能具有将卵菌效应蛋白输送到植物细胞质的功能（Tyler et al.

2006)。Joseph 等(2008)利用人类 β 肌动蛋白的 5 个结构基序(Ph1,C1,Ph2, Ad,C2)来划分盘基网柄菌的肌动蛋白或肌动蛋白相关蛋白,并对这些序列结构进行了分析,筛选出可能是盘基网柄菌的肌动蛋白必要的氨基酸。由图 2 可以看出,本研究获得的基序中仅有 motif2,motif3,Motif2 分别与 Joseph 等(2008)报道的基序 C1,Ad,C2 完全或部分重合,而 motif1,Motif1,Motif3 为本研究发现的保守基序。盘基网柄菌 32 条肌动蛋白保守基序 motif1,motif2,motif3 与动物界、变形虫界、植物界、真菌界、细菌界的肌动蛋白保守基序中 Motif1,Motif2,Motif3 进行比较发现,这 6 个基序有相互重合的部分,并且保守基序 motif1,Motif1, Motif3 在蛋白 Profilin-actin-VASP$_{202-244}$(PDB ID:3CHW)三维结构占重要位置。这表明,保守基序 motif1,Motif1,Motif3 可能是盘基网柄菌肌动蛋白重要的保守基序。因此,本研究结果可以看作是对以往研究的确认和延伸。对于保守基序 motif1,Motif1,Motif3 的进一步研究可以结合基因敲除、RNA 干扰等相应生物学实验,将可以更好验证这些基序在运动机理中的生物功能。

研究模式生物盘基网柄菌肌动蛋白对于揭示此类蛋白在盘基网柄菌及其他生物体内的生理过程的分子机制具有借鉴意义和参考价值。

References

Altschul, S. F., Gish, W., Miller, W., et al. 1990. Basic local alignment search tool. Journal of Molecular Biology, 215(3): 403 – 410.

Annesley, S., Fisher, P., 2009. *Dictyostelium discoideum*—a model for many reasons. Molecular and Cellular Biochemistry, 329(1 – 2): 73 – 91.

Baek, K., Liu, X., Ferron, F., et al. 2008. Modulation of actin structure and function by phosphorylation of Tyr – 53 and profilin binding. Proceedings of the National Academy of Sciences, 105(33): 11748 – 11753.

Bailey, T. L., Boden, M., Buske, F. A., et al. 2009. MEME SUITE: tools for motif discovery and searching. Nucleic Acids Research, 37(Web Server issue): W202 – 208.

Basu, S., Fey, P., Pandit, Y., et al. 2012. DictyBase 2013: integrating multiple Dictyostelid species. Nucleic Acids Research, 41(Database issue): D676 – 683.

Birch, P. R. J., Rehmany, A. P., Pritchard, L., et al. 2006. Trafficking arms: oomycete effectors enter host plant cells. Trends in Microbiology, 14(1): 8 –

11.

Condeelis, J., Hall, A., Bresnick, A., et al. 1988. Actin polymerization and pseudopod extension during amoeboid chemotaxis. Cell Motility and the Cytoskeleton, 10(1-2): 77-90.

DeLano, W. L., 2002. PyMOL molecular viewer. http://www.pymol.org/

Deng, H., Liu, C., Li, J., et al. 2007. The analysis of *Arabisopsis thaliana* guardcell-specific genes promoter motifs. Shengwu Jishu Tongbao, (3): 145-148; 151 (in Chinese).

Dominguez, R., Holmes, K. C. 2011. Actin structure and function. Annual Review of Biophysics, 40: 169-186.

Drouin, G., Moniz de Sa, M., Zuker, M. 1995. The *Giardia lamblia* actin gene and the phylogeny of eukaryotes. Journal of Molecular Evolution, 41(6): 841-849.

D'Haeseleer, P. 2006. What are DNA sequence motifs? Nature Biotechnology, 24(4): 423-425.

Eichinger, L., Pachebat, J. A., Glockner, G., et al. 2005. The genome of the social amoeba *Dictyostelium discoideum*. Nature, 435(7038): 43-57.

Frith, M. C., Saunders, N. F. W., Kobe, B., et al. 2008. Discovering sequence motifs with arbitrary insertions and deletions. Plos Computational Biology, 4(5): e1000071.

Guan, X. G., Su, W. H., Yi, F., et al. 2010. NPA motifs play a key role in plasma membrane targeting of aquaporin-4. IUBMB Life, 62(3): 222-226.

Hatayama, M., Aruga, J., 2010. Characterization of the tandem CWCH2 sequence motif: a hallmark of inter-zinc finger interactions. BMC Evolutionary Biology, 10(1): 1-18.

Joseph, J. M., Fey, P., Ramalingam, N., et al. 2008. The actinome of *Dictyostelium discoideum* in comparison to actins and actin-related proteins from other organisms. Plos One, 3(7): e2654.

Kapoor, P., Sahasrabuddhe, A. A., Kumar, A., et al. 2008. An unconventional form of actin in protozoan hemoflagellate, *Leishmania*. Journal of Biological Chemistry, 283(33): 22760-22773.

Keeling, P. J. 2001. Foraminifera and Cercozoa are related in actin phylogeny: two orphans find a home? Molecular Biology and Evolution, 18(8): 1551-1557.

Knecht, D. A., Cohen, S. M., Loomis, W. F., et al. 1986. Developmental regulation of *Dictyostelium discoideum* actin gene fusions carried on low-copy and high-copy transformation vectors. Molecular and Cellular Biology, 6(11): 3973 –3983.

Lahr, D. J. G., Grant, J., Nguyen, T., et al. 2011. Comprehensive phylogenetic reconstruction of amoebozoa based on concatenated analyses of SSU – rDNA and actin genes. Plos One, 6(7): e22780.

Loewy, A. G. 1952. An actomyosin-like substance from the plasmodium of a myxomycete. Journal of Cellular and Comparative Physiology, 40(1): 127 – 156.

Niu, T., Liu, X., Li, X., et al. 2013. Bioinformatic prediction and experimental validation of NDRG2 interaction partners. Zhongguo Shengwu Huaxue yu Fenzi Shengwuxue Bao, 29(2): 168 – 174(in Chinese).

Perry, S. V. 2003. When was actin first extracted from muscle? Journal of Muscle Research and Cell Motility, 24(8): 597 –599.

Raper, K. B., Rahn, A. W. 1984. The Dictyostelids. Books on Demand. Princeton: Princeton University Press. 1 –466.

Rehmany, A. P., Gordon, A., Rose, L. E., et al. 2005. Differential recognition of highly divergent downy mildew avirulence gene alleles by *RPP*1 resistance genes from two Arabidopsis lines. The Plant Cell, 17(6): 1839 –1850.

Spudich, J. A., Lord Wttao, K. 1974. Biochemical and Structural Studies of Actomyosin-like Proteins from Non-Muscle Cells: Isolation and characterization of myosin from amoebae of *Dictyostelium discoideum*. Journal of Biological Chemistry, 249(18): 6013 –6020.

Tamura, K., Peterson, D., Peterson, N., et al. 2011. MEGA5: Molecular evolutionary genetics analysis using Mmaximum likelihood, evolutionary distance, and maximum parsimony methods. Molecular Biology and Evolution, 36(5): 1 –24.

Thompson, J. D., Gibson, T. J., Plewniak, F., et al. 1997. The CLUSTAL_X windows interface: flexible strategies for multiple sequence alignment aided by quality analysis tools. Nucleic Acids Research, 25(24): 4876 –4882.

Tyler, B. M., Tripathy, S., Zhang, X., et al. 2006. *Phytophthora* genome sequences uncover evolutionary origins and mechanisms of pathogenesis. Science, 313(5791): 1261 –1266.

Uyemura, D. G., Brown, S. S, Spudich, J. A. 1978. Biochemical and structural characterization of actin from *Dictyostelium discoideum*. Journal of Biological Chemistry, 253(24): 9088 – 9096.

Vicker, M. G. 2002. F-actin assembly in Dictyostelium cell locomotion and shape oscillations propagates as a self-organized reaction-diffusion wave. FEBS Letters, 510(1 – 2): 5 – 9.

Yu, Q., Ma, L., Liu, L., et al. 2008. Primary analysis of host-targeting-motif harbored secreted proteins in genome of *Fusarium graminearum*. Shengwu Jishu Tongbao, (1): 160 – 165, 180(in Chinese).

邓荟芬,刘春林,李进,等. 2007. 拟南芥保卫细胞中特异性表达基因启动子片段基序分析. 生物技术通报,(3): 145 – 148, 151.

田李,陈捷胤,陈相永,等. 2011. 大丽轮枝菌(*Verticillium dahliae* VdLs.17)分泌组预测及分析. 中国农业科学,44(15): 3142 – 3153.

于钦亮,马莉,刘林,等. 2008. 禾谷镰刀菌基因组中含寄主靶向模体分泌蛋白功能的初步分析. 生物技术通报,(1): 160 – 165, 180.

Guang Li

Research Assistant, Beijing Proteome Research Center, China.

Research projects have been focused on comparative genomics of fungi, as well as on chromosomes evolution and development of mammalian.

Education

2007 – 2011; Jilin Agricultural University, Bachelor; 2011 – 2014; Yanbian University, Master.

Work experience

2014 – present; Research Assistant, Beijing Proteome Research Center, China.

Amplification and Sequencing of EF − 1α Region from *Didymium Squamulosum*

Shuyan Liu, Fengyun Zhao, Yu Li

Engineering Research Center of Chinese Ministry of Education for Edible and Medicinal Fungi, Jilin Agricultural University, Changchun, Jilin Province, China

Abstract: *Didymium squamulosum* is a widely distributed species throughout the world. To get the EF − 1α region from *D. squamulosum*, DNA was extracted from 3 − 5 adjacent sporophores by using 5% Chelex − 100. Semi-nested PCR was used for DNA amplification. Thirty-five cycles were conducted in a TC − 512 thermo cycler: 94°C for 30 s, 52°C for 1 min, 72°C for 1 min 30 s, with a final extension at 72°C for 10 min. PCR products were separated on 1.0% agarose gels, stained with ethidium bromide and viewed under UV light. The obtained 944 bp sequence was of 87% identity with that of *Physarum polycephalum*. This study not only enriched the EF − 1α genes sequences, but also provided evidence for the phylogeny study of myxomycetes.

Key words: Myxomycetes; elongation factor; molecular phylogeny

1 Introduction

Myxomycetes is unusual eukaryotic organism, and most of them like to live in moist forests. Its vegetative stage are plasmodia which can crawl and feed, and induce static reproductive sporophore or sclerotia in the case of starvation or lack of water (Martin et al. 2003).

In recent years, there are some reports about molecular phylogeny of myxomycetes based on the analysis of SSU rRNA and elongation factor EF − 1α gene (Fiore-Donno et al. 2005, 2008, 2013; Lundblad et al. 2004; Nandipati et al. 2012). However, there are only few sequences from few taxa available for the phylogenetic analysis. In order to enrich the number of EF − 1α gene sequences, we use *Didymium squamulosum* as an example to explore the optical reaction system and sequence it.

2 Materials and methods

2.1 Materials

Specimens were obtained from field-collection and stored in Herbarium of Mycology of Jilin Agricultural University (HMJAU).

2.2 Methods

2.2.1 DNA extractions

DNA was extracted from 3 − 5 adjacent sporophores by using 5% Chelex − 100 (Phillips et al. 2012). DNA was resuspended in 100 μL sterile water and stored at 4℃.

2.2.2 DNA amplification

The primers used in this research are shown in Table 1. DNA fragment was amplified by using semi-nested PCR. Thirty-five cycles were conducted in a TC − 512 Thermo Cycler. First PCR is as follows: 94℃ for 30 s, 52℃ for 1 min, 72℃ for 1 min 30 s, with a final extension at 72℃ for 10 min; second PCR: The annealing temperatures were 54°C. PCR products were separated on 1.0% agarose gels, stained with ethidium bromide and viewed under UV light.

3 Results and discussion

The EF − 1α sequence was obtained by nested PCR and sequenced. The fragment length was 994 bp (Fig. 1). It showed 87% identity to *Physarum polycephalum*.

Didymium squamulosum is a common species of Didymiaceae, Physarales of Myxomycetes in China, but there were no reports about its molecular phylogeny based on the analysis of EF − 1α genes. And now there is only one EF − 1α genes sequence of *D. squamulosum* stored on GenBank. Its length is only 362 bp. It is too short to be used in the phylogenetic analysis. This study not only enriched EF − 1α genes sequences, but also provided evidence for phylogeny of myxomycetes.

1   ACGAATGAGA GTCGAGCGTC GAGCCTGTTC GCCGATACGC AGATTGCTCT GTGGAAGTAG

61   GATACTGCCA AGTACTACAT CACCATCATT GATGCCCCCG GACATCGTGA CTTCATCAAG

121   AACATGATCA CTGGTACCTC CCAGGTATGT TTCTTTCGCG ATGCATTTTT ATATATGGTT

181   TTCCACCTTT GCATTTTATC TTTCACATAC ATGTTTAACA CATACCTGCG CATTTGCATT

241   GCATATACTT ATGGTTTATT CAAGGCTGAT GCCGCCGTGT TGGTCATTGC CTCCCCCACT

301   GGTGAGTTCG AGGCCGGTAT TGCCAAGAGC GGACAGACCC GTGAGCACGC TCTCCTTGCC

361   TTCACCCTCG GTGTGAGACA AATGATAGTT GCCATCAACA AGATGGACGA GAAGTCCGTC

421   AACTACGGCC AAGCCCGTTA TGACGAGATC GTCAAGGAGA CCTCCGCCTT CGTCAAGAAG

481   ATCGGATACA ACCCCGAGAA GATCAACTTC GTCCCCATCT CTGGATGGAA CGGAGACAAC

541   ATGTTGGAGA AGTCCCCCAA CATCCCATGG TACAAGGGAC CCACACTCCT CGAGGCCCTC

601   GATGCCATCA CCGAGCCCAA GCGCCCCAAC GACAAGCCCC TCCGTGTCCC CCTCCAGGAT

661   GTCTACAAGA TCGGAGGTAT TGGAACGGTT CCCGTCGGTC GTGTTGAGAC TGGTGTCCTC

721   AAGCCCAACA TGGTCGTCAC CTTCGCCCCC GGTAACTTGT CCACTGAGGT CAAGTCTGTT

781   GAGATGCATC ACGTCGCTCT CCCTGAGGCC ACCCCCGGAG ACAACGTTGG TTTCAACGTA

841   AAGAACTTGT CCGTCAAGGA TATCCGCCGT GGTATGGTCG CTGGTGACTC CAAGAACGAC

901   CCTCCCCGTG AGACTGAGTC CTTCACCGCC CAAGTCATCA TCCTCAACCA CCCCGGACAG

961   ATCCACGCCG GATATGCACC AGTGTGATGT GCGC

Fig. 1 The partial sequence of EF – 1α gene from *Didymium squamulosum*.

References

Fiore-Donno, A. M., Berney, C., Pawlowski, J. A. N., et al. 2005. Higher-order phylogeny of plasmodial slime molds (Myxogastria) based on elongation factor 1 – A and small subunit rRNA gene sequences. Journal of Eukaryotic Microbiology, 52(3): 201 – 210.

Fiore-Donno, A. M., Meyer, M., Baldauf, S. L., et al. 2008. Evolution of dark-spored Myxomycetes (slime-molds): Molecules versus morphology. Molecular Phylogenetics and Evolution, 46(3): 878 – 889.

Fiore-Donno, A. M., Clissmann, F., Meyer, M., et al. 2013. Two-gene phylogeny of bright-spored myxomycetes (slime moulds, superorder Lucisporidia). PLoS ONE, 8(5): e62586.

Lundblad, E. W., Einvik, C., Ronning, S., et al. 2004. Twelve Group I introns in the same pre-rRNA transcript of the myxomycete *Fuligo septica*: RNA processing and evolution. Molecular Biology Evolution, 21(7): 1283 – 1293.

Martín, M. P., Lado, C., Johansen, S. 2003. Primers are designed for amplification and direct sequencing of ITS region of rDNA from Myxomycetes. Mycologia, 95(3): 474 – 479.

Nandipati, S. C., Haugli, K., Coucheron, D. H., et al. 2012. Polyphyletic origin of the genus *Physarum* (Physarales, Myxomycetes) revealed by nuclear rDNA mini-chromosome analysis and group I intron synapomorphy. BMC Evolution Biology, 12: 166.

Phillips, K., McCallum, N., Welch, L. 2012. A comparison of methods for forensic DNA extraction: Chelex – 100 ® and the QIAGEN DNA Investigator Kit (manual and automated). Forensic Science International: Genetics, 6(2): 282 – 285.

Shuyan Liu

Doctor's tutor, Professor of Jilin Agricultural University.

Laboratory of Plant Pathology, College of Agronomy, Jilin Agricultural University, Changchun, China.

Education

1989.9 - 1993.7, Jilin Agricultural University, Plant Protection, Bachelor;

1994.9 - 1997.7, Jilin Agricultural University, Plant Pathology, Master;

1997.9 - 2000.6, Jilin Agricultural University, Crop Cultivation and Breeding, Ph. D.

Academic appointments

1993.7—1994.9, Shuangliao Forest Bureau, Jilin Province, technician;

1999.7—2001.12, Jilin Agricultural University, Plant Pathology, instructor;

2002.1—2007.12, Jilin Agricultural University, Plant Pathology, associate professor;

2008.1—present, Jilin Agricultural University, Plant Pathology, professor;

2005.1—2005.6, Mie University, Japan, visiting scholar;

2006.3—2007.3, Guelph University, Canada, visiting scholar;

2005.11—2005.12, Wageningen University, Newtherlang, visiting scholar.

Nuclear DNA Contents of Four Orders of Myxomycetes Collected in Jilin, China

Shu Li[1], Bao Qi[1], Wan Wang[1], Makoto Kakishima[1,2], Qi Wang[1], Yu Li[1]

1. Engineering Research Center of Chinese Ministry of Education for Edible and Medicinal Fungi, Jilin Agricultural University Changchun, China;
2. University of Tsukuba, Japan

Abstract: With special life cycle living as both protozoa and fungi, myxomycete was included into protozoa phylogenetically. However, their phylogenetic position and evolution are still not clarified in the morphologically different groups. Nuclear DNA content is significant in many fields of research including ecology, taxonomy and evolution, and it has been proved by many tests in plants and animals. This study is aimed to elucidate the relationships among the myxomycetes based on DNA content variation, combining with the morphological characteristics and living environment. The morphological classification relationships were recognized. Thirty samples belonging to four orders for the measurements of nuclear DNA content were mainly collected from Jilin Province in China. Nuclear DNA content can be measured quickly and simply by flow cytometry (BD Accuri ® C6) using the base unspecific intercalating fluorochrome propidium iodide (PI) with *Saccharomyces cerevisiae* as the primary internal standard. There was a small range of DNA fluorescence absorption value and spore size in dark-spore myxomycetes as Physarales and Stemonitales. By contrast, a bigger range of DNA content compared with genus in light-spore myxomycetes including Liceales and Trichiales was estimated by fluorescence absorption value. The intraspecific DNA contents were within the limits, and that was close to spore size which have small discrepancies among the samples from different places like *Lycogala epidendrum* collected from Jilin Province, Sichuan Province and

Tibet Autonomous Region. It was suggested that DNA content was related with the differentiation of spore size. And DNA content may contribute to interspecific morphological divergence within genus.

Key words: Myxomycetes; DNA content; flow cytometry; *Lycogala epidendrum*

Resource Allocation and Morphogenesis during Fructification in Myxomycetes

Qian Li, Shuzhen Yan, Shuanglin Chen

College of Life Sciences, Nanjing Normal University, Nanjing, Jiangsu, China

Abstract: Physarida is the biggest order in the class Myxogastria. It mainly includes two families: Physaridae and Didymiidae. In order to explore the effects of internal transcribed spacers region of ribosomal DNA (rDNA ITS) on the phylogenetic analysis of Myxogastria, we analyzed the primary structure and secondary structure of ITS, and the phylogenetic relationships in Physarida.

The universal primers PHYS4 and PHYS5 were used to amplify and sequence the rDNA ITS sequences from eight species of five genera in the order Physarida. The obtained sequences were combined with known sequences in GenBank to construct a phylogenetic tree using both Maximum Likelihood (ML) and Bayesian Inference (BI) methods. The varieties for the rDNA ITS of different Physarida species were observed both on the base composition and length. The range of length is 777 – 1445 bp, and G + C mol % is between 53.4% and 61.9%. Physarida and Stemonitida clustered for two distinct branches. On the branch of Physarida, Physaridae and Didymiidae were divided into two separate branches respectively. The results supported those taxonomic viewpoints which make a distinction between the two families based on capillitium with or without calcareous granules in the morphology. The samples of *Didymium squamulosum* from different geographic regions are composed of three branches, which again confirmed that this morphospecies is consisted of a biological species complex with different geographical origins, reproductive incompatibility and genetic variations.

The RNA secondary structures of ITS were calculated and drawn using the program RNA structure. The results showed that the ITS1 secondary structures in most tested Physarida species are different, in which only 24 bp sequences are conservative. The base composition are A (G) C (U) C CGC (U) ACUGGUGAACCUGCGGGU(C). There is a common stable helix structure in most tested Physarida species. Covariation happened in three positions that maintain the complementary base pairing. The stable helix structure plays an important role in the processing of mature rRNA. The 5.8S secondary structures in different species were similar to each other, which are composed of four helixes with two main types. The ITS2 secondary structure was constructed based on the 5.8S rRNA – 28S rRNA interaction. A closed multibranch loop and at least four major helixes were predicted with the helix IV being relatively conservative. The ITS2 secondary structure of Physarida species can form a core around with several regions in pairs, and is a closed multibranch loop. At least four major helixes were predicted with all tested Physarida species, among them the helix IV being relatively conservative, and there are no branch ring which will influence the stability of secondary structure. The varieties of length in different species are also smaller. The secondary structures of ITS regions were more conserved than their nucleotide sequences. Therefore, further analysis of ITS structures will be helpful for the understanding of the relationships between the ITS structure and molecular evolution of Physarida species.

Key words: Physarida; rDNA ITS; molecular evolution; sequence; secondary structure

Shuanglin Chen

Professor, College of Life Sciences, Nanjing Normal University, Jiangsu, China.

Work experiences

Professor, Nanjing Normal University, China.

Major research area

Research projects have been focused on morphology, taxonomy, diversity and molecular biology of myxomycetes in China, as well as on diversity of plant endophytes, biological characteristics, activities and early development of natural products.

Major researches

60 papers, 3 books.

What An Intron May Tell:
An Analysis of Two Markers in *Meriderma* spp. (Stemonitales)

Martin Schnittler[1], **Eva Heinrich**[1], **Alexander Kettler**[1], **Thomas Sura**[1], **Yuri K. Novozhilov**[2]

1. Institute of Botany and Landscape Ecology, Ernst Moritz Arndt University Greifswald, Grimmer Str. 88, D-17487 Greifswald, Germany;
2. V. L. Komarov Botanical Institute of the Russian Academy of Sciences, Prof. Popov St. 2, 197376 St. Petersburg, Russia

Abstract: Molecular phylogenies revealed the genus *Meriderma*, traditionally known as the *Lamproderma atrosporum* group, to be a basal clade within dark-spored myxomycetes. Recently, the pioneering studies of Poulain, Meyer & Bozonnet led to the differentiation of several taxa, which encouraged us to start a molecular investigation, including specimens from the French and German Alps, the Caucasus, and the Rocky Mountains. To our expectation, by studying partial SSU sequences we found more than 50 genotypes within ca. 200 investigated specimens. Even more interesting was the first part of the protein elongation factor EF1-α which we sequenced as a second marker. In this gene, the protein-coding sequence is interrupted by an extremely variable spliceosomal intron.

More than three thirds of the 70 specimens investigated for this marker were heterozygous in the intron section; substitution but especially deletions complicated sequencing. A major SSU genotype of *M. aggregatum* shows a unique intron which is nearly inaccessible for sequencing, consisting of a stretch of >14 subsequent C followed by a dinucleotide SSR with the motif CA. In contrast, genotypes of specimens assigned to *M. spinulisporum*, *M. echinulatum*, and *M. carestiae* showed sequences rich in T, which are typical for splicesomal introns.

The high proportion of specimens with heterozygosities in the intron points to frequent sexual reproduction. However, in several cases we found the same heterozygosity pattern in specimens collected at different locations, even at different mountain ranges. This lets dispersal via microcysts and (or) amoebae appear to be unlikely: according to textbook on (sexual) life cycle these propagules should be haploid, thus transporting only one allele. An alternative explanation would be to assume an optional asexual life cycle with diploid amoebal stages and spores, which could transport both alleles. The occasional observation of "macrosporic" forms in several taxa could be a hint on the existence of such an apomictic life cycle, which was postulated already by earlier studies with cultivated members of the Physarales.

Key words: 18S rRNA gene; intron; protein elongation factor EF1 − α

The Genus *Alwisia* (Myxomycetes) Revalidated, with Three Species New to Science

Dmitry Leontyev[1], Martin Schnittler[2], Steven L. Stephenson[3], Gabriel Moreno[4], David W. Mitchell[5], Carlos Rojas[6]

1. Kharkiv State Zooveterinary Academy, Akademichna str. 1, Kharkiv 621343, Ukraine;
2. Institute of Botany and Landscape Ecology, Ernst Moritz Arndt University Greifswald, Grimmer Str. 88, D – 17487 Greifswald, Germany;
3. University of Arkansas, Fayetteville, Arkansas 72701, USA;
4. Universidad de Alcalá, Alcalá de Henares, España 28805, Madrid, Spain;
5. Walton Cottage, Upper Hartfield, East Sussex, TN74AN, England, UK;
6. Universidad de Costa Rica, San Pedro de Montes de Oca11501, Costa Rica

Abstract: Based on morphological investigations and 18S rRNA phylogeny, we revalidate the formerly monotypic genus *Alwisia*. The monotypic genus *Alwisia* Berk & Broome (Reticulariaceae), described in 1873, was a few decades later united with *Tubifera*, and the new combination *Tubifera bombarda* (Berk & Broome) G. W. Martin was proposed for its single species. However, 18S rRNA sequences revealed that *T. bombarda* forms a separate clade within the family Reticulariaceae which should be recognized as a separate genus under its original name *Alwisia*.

The same locus of 18S rRNA gene was studied in several unidentified collections from Costa Rica, Australia and Tasmania. Obtained sequences appeared to be related, but yet different from *A. bombarda*, and seem to be identical within the specimens of one morphotype. Studied collections differ from *A. bombarda* in morphological characters as well. Therefore we described them as three new species within the revalidated genus.

All new species have subspherical sporothecae, against fusiform ones in *A. bombarda*. Among them, *Alwisia morula* is characterized by erect and branched individual stalks, while *Alwisia repens* has procumbent stalks and iridescent peridium. Both *A. morula* and *A. repens* completely lack a capillitium, while the third new species, *Alwisia lloydiae*, possesses a tubular capillitium ornamented with globular warts.

Capillitial structures of *Alwisia bombarda* and *A. lloydiae* occur inside separate sporocarps with intact walls. Therefore they do not represent remnants of confluent peridia and thus do not correspond to the definition of a pseudocapillitium, but should be seen as a true capillitium. Capillitial threads in *Alwisia* resemble those found in the genera *Dianema* and *Lycogala*, thus providing a new argument for a close relationship between Reticulariaceae and Dianemataceae and also for considering the tubular threads of *Lycogala* as a true capillitium.

Key words: 18S rRNA gene; *Tubifera bombarda*; *Alwisia bombarda*; *A. Morula*; *A. Repens*; *A. lloydiae*

New Insights into the *Tubifera ferruginosa*-Complex

Dmitry Leontyev[1], Martin Schnittler[2], Steven L. Stephenson[3]

1. Kharkiv State Zooveterinary Academy, Akademichna str. 1, Kharkiv 621343, Ukraine;
2. Institute of Botany and Landscape Ecology, Ernst Moritz Arndt University Greifswald, Grimmer Str. 88, D – 17487 Greifswald, Germany;
3. University of Arkansas, Fayetteville, Arkansas 72701, USA

Abstract: A phylogeny based on partial 18S rRNA gene sequences revealed that *Tubifera ferruginosa* (Batsch) J. F. Gmel. is actually a complex of seven different species, showing a high degree of sequence dissimilarity especially in the variable helices. Among them, *T. applanata* Leontyev & Fefelov and *T. dudkae* ad int. (comb. nov. pro *Reticularia dudkae* Leontyev & G. Moreno) were recently described and now confirmed by species-specific 18S rRNA sequences which are identical for specimens from different geographic origins.

Within *T. ferruginosa* sensu strictu we recognize ssp. *ferruginosa* ad int. and ssp. *acutissima* ad int., differing constantly by hemispherical to obtusely conical vs. acute conical sporothecae with subulate apices, respectively.

The remaining taxa are described as new to science; their 18S rRNA phylogeny corresponds to morphological characters like the structure of sporothecal tips, color of immature fructifications and ultrastructure of the inner peridial surface. *T. montana* ad int. develops from orange-red plasmodia the sporothecae with strongly accreted tips and peridia iridescing by golden to pinkish hues; spores are considerably larger than in *T. ferruginosa*.

The large pseudoaethalia (3 – 12 cm long) of *T. magna* ad int. have flattened sporothecal tips resembling those of *T. applanata*. However, in *T. applanata* immature fructifications are flesh-colored to salmon, and sporothecal tips are isodiametric and roughly hexagonal, while *T. magna* appears with pink colors, and its sporothecal tips are elongate and variable in shape.

T. pseudomicrosperma ad int. differs from *T. ferruginosa* by its thick, black hypothallus and *beige peridium*, and from *T. microsperma* by much larger pseudoaethalia, prostrate hypothallus and even smaller spores. Its peridium shows small rimmed craters indistinguishable in phase contrast LM, while in *T. microsperma* these craters are larger and clearly visible.

T. corymbosa ad int. possesses small spherical sporothecae at the base of the pseudoaethalium resembling those of *T. dimorphotheca*, but lacks the prominent hypothallic stalk of the latter species and has a peridium, iridescent in blue and green tints, with *a metallic, silvery or golden luster on the sporothecal tips*.

All taxa of the *T. ferruginosa*-complex lack a capillitium; structures described by Nannenga-Bremekamp (1961, 1991) with LM and SEM were found to be fungal hyphae feeding on spores.

Key words: 18S rRNA gene; species differentiation; capillitium

第六部分
Part VI

生物学
Biology

Application of 3D Imaging of Light and Electron Microscopy in Studying Myxomycetes

Yuka Yajima

Graduate School of Medicine, Kyoto University 606-8501, Japan

Abstract: Developing and mature myxomycete fruiting bodies have complicated structures which are usually too thick and opaque to observe the inside under light microscope (LM), and too big to understand the arrangement of their complex by transmission electron microscope (TEM). To understand the structure inside of myxomycetes, the author combined the conventional serial sectioning technique and the computer-based three-dimensional (3D) imaging technique.

3D imaging is an effective technique to define the structure of organisms, and the use of computers to aid in the reconstruction and segmentation of the large data files of the imaging is now widely available to biological researchers. We have several high-end electron microscopes to acquire the raw data of 3D image nowadays, and yet general-use computers are still not up to the task to reconstruct and segment the entire structure of myxomycetes in electron microscopic level, since its big and complex structure generates too large data files to process.

Here, the author presents some application examples of 3D imaging for myxomycete research. The author acquired the dataset by the conventional serial sectioning technique, reconstructed and segmented by a general-use computer. The correlation between 2D and 3D imaging of light and electron microscopy links the overview and orientation of the complex structure inside the myxomycetes, and detailed localization and correlation of subcellular structures. Although the 3D imaging is a technically demanding and time-consuming technique, it has the potential to provide new insights about the morphogenesis and morphological characteristics of myxomycetes.

Key words: Three-dimensional imaging; serial sections; reconstruction; segmentation; TEM; LM

一种准确测定黏菌原质团原生质流流速的方法

王晓丽[1,2]，李晨[2]，李艳双[1]，李玉[1]

1 吉林农业大学食药用菌教育部工程研究中心，吉林长春；
2 吉林农业大学农学院，吉林长春

摘要：在摄录的一段黏菌显型原质团原生质流的视频中，确定流经的明显的液泡，做一系列视频截图，在截图中确定液泡的周长及中心点，重叠所做截图，得到气泡的运动轨迹。通过显示的相邻气泡中心点的距离，可以得到总距离，最终得到这个液泡在这一段时间的流速。视频截图时间越小得到的流速越准确。测定若干个液泡的流速，其平均值可以代表此区段内原生质流的流速。

关键词：黏菌；显型原质团；原生质流；液泡

An Accurate Method to Measure Velocity of Protoplasm Streaming in Myxomycetes Plasmodium

Xiaoli Wang [1,2], Chen Li [2], Yanshuang Li [1], Yu Li [1]

1. Engineering Research Center of Chinese Ministry of Education for Edible and Medicinal Fungi, Jilin Agricultural University, Changchun, Jilin Province, China
2. College of Agronomy, Jilin Agricultural University, Changchun, Jilin Province, China

Abstract: In recording a streaming video of myxomycete phaneroplasmodium, we determined clear vacuoles through the channel, did a series of video capture, and determined the circumference of the vacuole and center in the screenshot. By overlapping of the screenshots, the trajectories of bubbles can be obtained. By displaying the center distance of adjacent bubbles, we were able to get the total distance, finally obtained the velocity of the vacuoles in a period of time. The less time of the video capture is, the more accurate the velocity is. Determining the average flow velocity of a number of vacuoles can represent the flow rate of the plasma flowing in that section.

Key words: Myxomycetes; phaneroplasmodium; protoplasm streaming; vesicles

一、概述

黏菌原质团是黏菌的无性生长阶段（Alexopoulos 1960），是大型的、多核的非细胞结构。在原质团中物质运输的主要方式是原生质流，前人对原生质流的研究表明，原生质朝向原质团移动方向的流动时间不一定长于反向流动时间（Kamiya 1950），原生质向前流动大约 1.5～3 min 就要进行往返运动（Wohlfarth-Bottermann & Fleischer 1976）。我们以淡黄绒泡菌的显型原质团为材料对原生质流的研究得出如下结论，原生质流动方向总体是原质团运动方向，反向原生质流的出现是因为原质团扇面前端尚未分化出菌脉通道，以及菌脉中凝胶态物质阻塞通道所致，这也正是原质团前缘形成扇面的原因。凝胶态物质向溶胶态转化是原生质流的最终动力（吕冬霞等 2012）。我们在此研究的基础上，找到一种可以准确测定原生质流流速的方法。此种方法也可以应用到其他较小的细胞中研究胞质环流。

二、材料和方法

（一）材料

淡黄绒泡菌 *Physarum mulleum* 显型原质团。

（二）方法

在淡黄绒泡菌显型原质团的原生质流中存在一些明显的液泡，这些是测定原生质流速的标志物，但首先需假定这些液泡的流动速度同其周围的原生质流的流速相同。首先摄录一段淡黄绒泡菌显型原质团某一菌脉区段的原生质流动视频，在视频中找到流经的明显的 9 个气泡（分别标记成 A, B, C, D, E, F, G,

H,I),求出9个气泡的平均流经速度,即可以认为是这段时间内此段菌脉中原生质流的流速。具体方法以 A 气泡为例:从 A 气泡出现在视野中一刻起到流出视野外的时间区段中,进行截图处理。每秒平均2帧截图,在视频中的 1′51″气泡进入视野,在 1′58″流出,共获得截图 14 幅,时间分别在 1′51″,1′51.5″,…,1′57″,1′57.5″。截图的时间长为 6.5 s。在每幅截图中使用 motic 中 USB2.0 camera 的 MIPlus 软件确定气泡的周长及中心点,气泡有一定的体积并且在运动过程中形状会发生变化,气泡的运动距离以中心点为准。在 Photoshop 中分别重叠 14 幅截图,使 14 个截图中的气泡出现在同一图上,根据 MIPlus 软件显示的 2 个气泡中心点的距离,计算最近 2 点的速度;再计算 A 气泡在 6.5 s 内的速度。9 个气泡速度的平均值代表了这个菌脉中的原生质流在某一时段的流动速度。并从曲线上可以看出流动方向以及持续的时间。为了保证图片的美观及可比性,保证标记物从视野的左下角进入,右上角流出。

三、结果

图1 是 A 气泡在 1′52″时的截图(图 1a)及根据 MIPlus 软件确定的周长及中心点(图 1b),以此方法确定 A 气泡 14 个截图中气泡的周长及中心点。图 2a 是 A 气泡在 Photoshop 把 14 个不同截图中的气泡体现在同一图中的 A 气泡在此视野中流经途径及中心点的连线,在 MIPlus 中可以看到相邻气泡中心点的距离,以此计算相邻气泡间的速度及 6.5 s 内的速度。表 1 是 A 气泡从 1′51″ - 1′57.5″ s13 个时间区段的路程及平均速度,分段速度从 144.4 ~ 213.2 μm/s,平均速度为 174.9 μm/s。图 2b - d 是 B 气泡在 3′48.5″ - 4′34″时间区段内的流经图及气泡中心点连线,共做了 92 个截图,在 3′56.5″ - 4′19.5″的时间区间内仅做了 2 个截图,其他时间为每秒 2 幅截图,这是根据气泡流动的快慢决定的,在 3′56.5″ - 4′19.5″这个时间区段内,气泡移动极其缓慢,仅仅移动了 22.5 μm,很难在 23 s 做出 46 个截图,像其他时间段一样。B 气泡的流经速度是 9 个气泡中速度最慢

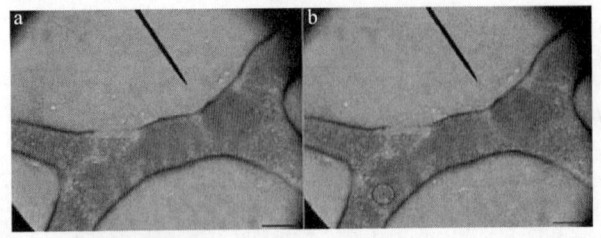

图 1 A 气泡在 1′52″的原始截图及根据 MIPlus 软件所做的气泡周长及中心点
Fig. 1 The original screenshot of bubble A at 1′52″ and determining the circumference and center point by MIPlus software.

的一个,仅为 25.83 μm/s。9 个气泡中,速度最快的是 D 气泡的 289.05 μm/s,9 个气泡的平均速度为 190.01 μm/s(表 2)。我们可以认为,这个速度是这段菌脉在这个时间区段内的原生质的流动速度。

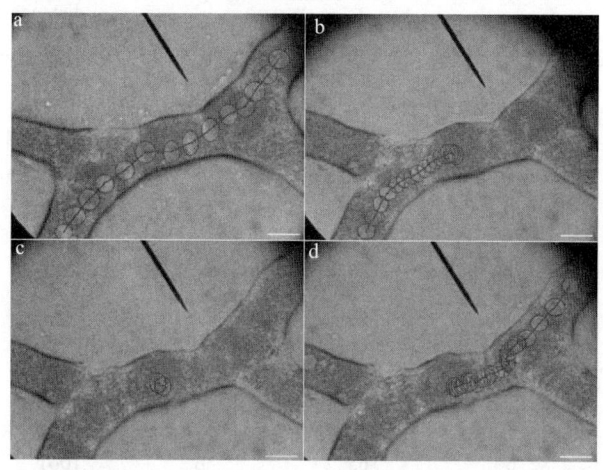

图 2 淡黄绒泡菌原质团某一菌脉中 A,B 气泡流经图及连接曲线

a,A 气泡流经图及连接折线;b-d,B 气泡流经图及连接折线,b, 3′48.5″-3′56.5″, c, 3′56.5″-4′19.5″, d, 4′19.5″-4′34″。

Fig. 2 Trajectory and connected curve of bubble A and B in vein of *P. melleum*.

a, Trajectory and curve of A; b—d: Trajectory and curve of B.

从图 2 中可以看出,如果不同气泡所做截图的时间区段相同,如 A 气泡和 B 气泡在 3′48.5″-3′56.5″和 4′19.5″-4′34″这两个时间区段的截图都是 2 幅/s,从图上可以直观看出气泡的流动快慢,气泡密集代表速度缓慢,如果气泡间距大则说明气泡流动较快。B 气泡在 3′48.5″-4′34″区段内的中间,即接近 3′56.5″-4′19.5″这个区段内气泡密集,说明气泡流经较慢,并在 3′56.5″-4′19.5″区段内仅仅移动 22.5 μm。而在进入视野和即将出去时间段内流动较快,气泡间距较大。

表 1 淡黄绒泡菌某一菌脉 A 气泡 6.5″流经的路程及速度
Table 1 Journey and velocity in 6.5″ of bubble A

时间区段 Time segment	分段路程/μm Piecewise journey	分段速度 /(μm·s⁻¹) Segmentation speed	经历时间/s Through time	累计路程/μm Cumulative distance	累计平均速度/(μm·s⁻¹) Cumulative average speed
1′51″-1′51.5″	76.4	152.8	0.5	76.4	152.8
1′51.5″-1′52″	85.7	171.4	1	162.1	162.1
1′52″-1′52.5″	88.6	177.2	1.5	250.7	167.1

续表

时间区段 Time segment	分段路程/μm Piecewise journey	分段速度 /(μm·s⁻¹) Segmentation speed	经历时间/s Through time	累计路程/μm Cumulative distance	累计平均速度/(μm·s⁻¹) Cumulative average speed
1′52.5″ – 1′53″	96.5	193	2	347.2	173.6
1′53″ – 1′53.5″	106.0	212	2.5	453.2	181.3
1′53.5″ – 1′54″	106.6	213.2	3	559.8	186.6
1′54″ – 1′54.5″	82.5	165	3.5	642.3	183.5
1′54.5″ – 1′55″	93.1	186.2	4	735.4	183.9
1′55″ – 1′55.5″	93.4	186.8	4.5	828.8	184.2
1′51.5″ – 1′56″	77.7	155.4	5	906.5	181.3
1′56″ – 1′56.5″	72.2	144.4	5.5	978.7	177.9
1′56.5″ – 1′57″	82.6	165.2	6	1061.3	176.9
1′57″ – 1′57.5″	75.9	151.8	6.5	1137.2	174.9

表 2 淡黄绒泡菌某一菌脉 9 个气泡流速测定
Table 2 The velocity measurement of 9 bubbles in same vein of *P. melleum*

气泡序号 No. of bubble	入出时间 Time segment	截图数量 Screenshot number	全程长度/μm Whole length	总观察时间 Total time	气泡平均速度 Average velocity/(μm·s⁻¹)
A	1′51″ – 1′57.5″	14	1137.18	6.5″	174.95
B	3′48.5″ – 4′34″	92	1175.3	45.5″	25.83
C	6′02.5″ – 6′11″	18	1241.6	8.5″	146.07
D	6′19.5″ – 6′23.5″	9	1156.2	4″	289.05
E	6′22″ – 6′27″	11	1274.1	5″	254.82
F	6′53″ – 7′07.5″	30	1153.7	14.5″	79.57
G	7′35″ – 7′39″	9	1122.0	4″	280.5
H	7′39.5″ – 7′45.5″	13	1050.9	6″	233.53
I	7′40.5″ – 7′45.5″	11	1129	5″	225.8
总平均速度 Total average velocity					190.01

四、讨论

细胞内的胞质环流及黏菌原质团中的原生质流是不同类型生物体的同一生

物学事件。胞质环流对于细胞的营养代谢具有重要作用,能够不断地分配各种营养物质,使它们在细胞内均匀分布,并且排出代谢废物。原生质流在原质团中也有着相同的作用。但对于绝大多数的动植物细胞来说,目前的实验技术很难准确测定胞质环流的流速。而能够观察胞质环流的细胞仅限于绿色植物的叶肉细胞,因其内部有叶绿体可以作为观察胞质环流的标志物。黏菌显型原质团的巨大性,使其在普通光学显微镜下极容易观察到原生质流,并且菌脉中流动的液泡及颗粒状物质都能作为研究原生质流的标记物质。

  本方法中,以显型原质团原生质流中的流动液泡作为标记物质,利用软件确定液泡的周长及中心点,以中心点的移动速率代表液泡的流动速率,以液泡代表周围原生质的流动速率。进行系列等时间段截图,确定两点之间的距离,计算速率。截图间距时间越短,最终的结果越接近于原生质实际的流速,标记物体越多,得出的平均值也越接近原生质流流动速率的实际值。一般研究表明,藻类胞质环流的流速是 10 ~ 50 μm/s,普通植物细胞为每秒几微米到几十微米(Shimmen et al.1982)。前人对黏菌多头绒泡菌的原生质流的研究得出的结论是原生质流流速为 1.35 mm/s(Alexopoulos et al.1996)。通过我们对淡黄绒泡菌原质团原生质流流速的精确测定得出的结论为 190 μm/s,比前人的估算要小得多。

  黏菌的相对类群较小,研究的也比较少,但随着对黏菌研究的不断深入,黏菌很可能成为许多研究的模式材料。最初人们只认识到黏菌显型原质团是研究细胞核周期生物学事件的模式材料,这主要是因为原质团内部存在大量同步有丝分裂的细胞核。而黏菌的黏变形体,其特点类似于人体内的白细胞、癌细胞及参加伤口愈合的细胞,尤其是黏变形体的不死性、可移动性、失去分裂上限,这些也使黏变形体可能成为研究如何控制癌细胞的替代材料。

参考文献

[1] Alexopoulos, C. J. 1960. Gross morphology of the plasmodium and its possible significance in the relationships among the myxomycetes. Mycologia, 52: 1 - 20.

[2] Alexopoulos, C. J., Mims, C. W., Blackwell, M. 1996. Introductory mycology. 4th ed. 788.

[3] Kamiya. 1950. The protoplasmic flow of the myxomycete plasmodium as revealed by a volumetric analysis. Protoplasma, 39:344 - 357.

[4] Shimmen, T., Tazawa, M. 1982. Cytoplasmic streaming in the cell model of Nitella. Protoplasma, 112: 101 - 106.

[5] Wohlfarth-Bottermann, K. E., Fleischer, M. 1976. Cycling aggregation patterns of cytoplasmic F-actin coordinated with oscillating tension force generation. Cell Tissue Research, 165: 327 – 344.

[6] 吕冬霞,蔚荣海,李景鹏,等. 2012. 淡黄绒泡菌和全白绒泡菌原生质流向的观察. 菌物学报,31(6):952 – 955.

Xiaoli Wang

Master's tutor, Professor of Jilin Agricultural University, Jilin, China.

Prof. Wang graduated from Jilin Agricultural University in 1987, received a bachelor's degree in agronomy, a master's degree in 1990, and a doctor's degree in 2004. She has been engaged in teaching and research work about plant genetics and breeding. In 2001, under the guidance of Professor Yu Li, she began to study cell biology of true slime mold, published about 20 relevant papers in more than 10 years, and won two prizes of natural science of Jilin Province.

A Comparative Study on the Developmental Processes of the Family Physaridae in the Pure Culture

Wei Tao, Shuzhen Yan, Shuanglin Chen

College of Life Sciences, Nanjing Normal University, Nanjing, Jiangsu, China

Abstract: Myxogastria (slime molds) are unique eukaryotic microorganisms with both characteristics of fungi and amoebae. Different development processes have been found in different species of Myxogastria. In this research, a comparative study on the developmental processes of the family Physaridae was conducted. Four myxogastrian species in the family Physaridae were purely cultured, and their life cycles were observed. They were *Physarum flavicomum*, *Physarum melleum*, *Physarum nutans* and *Physarella oblonga*. Sterile culture techniques including hanging drop culture and oats-agar culture, and some observation technologies for microstructures were used in order to study and compare the differences of the life history of Physaridae. The results showed that spores of these four myxogastrian species germinated all by V-shape split, and then produced one or more myxamoebae. Spores germination was observed within 2 d after inoculation, but germination time was slightly different among four Physaridae species in the same condition. Spores of *Physarum flavicomum* took up 5 h for germination to occur; however spores of *Physarum nutans* took up 1 d, spores of *Physarum melleum* took up 36 h, and spores of *Physarella oblonga* took up 2 d. A typical V-shape split was observed in the spores of *Physarella oblonga* and *Physarum flavicomum*, the length of the crack accounted for almost two thirds of the spore's diameter; however only a narrow gap was formed in the spores of *Physarum melleum* and *Physarum nutans*. The myxamoeba of *Physarum melleum* and *Physarum melleum* changed into a swarm cell with two flagella, whereas in the same conditions swarm cell was not found in *Physarum flavicomum* or *Physarum nutans*. It was observed that several young plasmodia of *Physarum melleum* were found on the oat agar in 9 d after spore inoculation; however young plasmodia of *Physarella oblonga* and *Physarum flavicomum* were observed in 12 d after spore inoculation. It required more time to observe young plasmodia of *Physarum nutans*, which was about 16 d. Young plasmodia of *Physarella oblonga* and *Physarum nutans* required about 10 d to grow all over the plate, whereas young plasmodia of *Physarum melleum* and *Physarum flavicomum* needed 18 d to become maturity after inoculation. It is proposed that the plasmodia of different myxogastrian species had different growth rates. Plasmodia of these four myxogastrian species had obvious differences in colour and veins. These four myxogastrian species could produce mature fruiting bodies on oat agar medium, which sporulation all needed light stimulation. The plasmodia of *Physarella oblonga* and *Physarum melleum* needed 2 – 3 d to produce young fruiting bodies. Furthermore, the plasmodia of *Physarum flavicomum* transformed to young fruiting bodies in 7 d, whereas to *Physarum nutans*, the time from maturity of plasmodia to sporulation needed 13 d in the same culture condition.

Key words: Myxogastria; spores germination; plasmodia; sporulation

Some Hypotheses about Lepidoderma

Renato Cainelli

Via Locchi 42 34123 Trieste, Italy

Abstract: Slime molds' sporangia are not the result of a predetermined assembly plan that ensures morphological consistency, as in multicellular organisms, but rather the result of a series of coordinated processes. Some of these processes may be affected by environmental conditions causing variations that can make sometimes impossible the correct determinations of the species or even the genus according to macroscopic characteristics. The author will propose some simple interpretative keys which may be useful in determining the species of the genus Lepidoderma. This work is based on the direct observation of the phenomenon of recrystallization of the lime that can affect many Physarales and on some simple assumptions on the external structure of the peridium.

Key words: Lepidoderma; recrystallization; external structure of the peridium

Distribution and the Food Resource Preference of Protostelids in Sugadaira Highlands, Nagano, Japan

Y. Iwamoto[1], Y. Degawa[1], J. Matsumoto[2]

1. University of Tsukuba, Japan;
2. Fukui Botanical Garden, Japan

Abstract: In Japan, the biodiversity of true slime molds and dictyostelids is well investigated; on the other hand, such a study on protostelids is extremely deficient. There is not yet any comprehensive monograph of Japanese protostelids (except a few partial records of some species: four spp. in Moore et al. 1995). It has been known that some species of protostelids, for example *Protostelium mycophagum*, have the preference of food resource. But their preferences have not been precisely examined from ecological or physiological view point. Then, in order to start the floristic survey of Japanese protostelids at first, we try to clarify the relationship between the distribution in micro-scale and the preference for food resources of protostelids, using baiting method. The study was performed in Sugadaira Montane Research Center, Sugadaira Highlands (c. 35 ha, c. 1360 m alt.), Nagano, Japan in temperate climate zone. As baits, sterilized three kinds of substrates (2 mm squared cut culm of *Miscanthus sinensis*, bark of *Pinus densiflora* and bark of *Quercus crispula*) were placed on the litter of four sites of different vegetation (*M. sinensis* grassland, *P. densiflora* forest, *Quercus* forest, Daimyoujin-no-taki waterfall in *Q. crispula* forest in Sugadaira Montane Research Center). After 3 weeks samples were recovered, brought back to the laboratory and soaked in sterilized distilled water. Eight pieces of each sample were inoculated onto wMYA (weak Malt extract and Yeast extract Agar) plates and observed under light microscope with long focus lens for 2 weeks. We counted the number of the pieces on which protostelids appeared, and calculated their frequencies. As a crude culture, the spore was isolated from recognized fruiting body and cultured with two isolates of bacteria obtained from each site by dilution plate method. For judging whether the preferences against food resources (certain isolates of bacteria) exist or not, the presence of spore germination and the formation of fructification under cultures were checked. These processes were performed two times in August and October. Protostelid fruiting bodies appeared on 30 baits from among 196 pieces. Four isolates of two-membered cultures were established among 30 crude cultures. Two isolates of them were identified as *Protostelium arachisporum*, and one was identified as *Schizoplasmodiopsis vulgare* based on their morphological characteristics and life cycles. The other one isolates could not be identified. *P. arachisporum* and *S. vulgare* were newly recorded from Japan. The total frequency of protostelids was highest on *M. sinensis* grassland. But they did not always prefer the isolate of bacteria from *M. sinensis* grassland; in another word, the preference to the food could not be clearly detected. Thus, in *M. sinensis* grassland, high frequency of protostelids may be caused not by the preference of the food resources but by the environmental factor such as instability of the temperature and humidity. In general, the frequency of protostelids may be higher, when the baits are placed on the site where the substrates do not originally exist.

Key words: Prototelids; food resource preference

Dr. Jun Matsumoto

Graduate university and major

Graduate School of Science, Hiroshima University, Department of Biological Science, Japan.

Work organization

Director at Fukui Botanical Garden.

Major research area

Taxonomy on Myxomycetes; Botany (taxonomy, biodiversity).

Major researches (papers, books or others)

6 papers.

Study on Isozyme in Different Ontogenetic Stages of *Didymium iridis*

Shicui Jiang, Bo Zhang, Yu Li

Engineering Research Center of Chinese Ministry of Education for Edible and Medicinal Fungi, Jilin Agricultural University, Changchun, China

Abstract: The spore to spore cultures of *Didymium iridis* were completed in our lab. And details of its life cycle are provided (including those from spores, myxoamoeba, zygote, young plasmodium, mature plasmodium, young fruiting bodies to mature fruiting bodies). In this research using the four different ontogenetic stages of *Didymium iridis* (mature plasmodium, split phase of plasmodium, the young fruiting bodies and mature plasmodium) as material to extract the isozyme, and using PAGE (polyacrylamide gel electrophoresis) to analyze, we showed that zymogram bands of the mature plasmodium were in accordance with the split phase of plasmodium, and zymogram bands of the young fruiting bodies were in accordance with the mature fruiting bodies. But there were distinctive bands between the mature plasmodium, the split phase of plasmodium and young fruiting bodies, mature fruiting bodies. So, isozyme variation in different ontogenetic stages could be used as an indicator of physiological and genetic identification of *Didymium iridis*.

Key words: Isozyme; ontogenetic stage; myxomycetes; life cycle

Fatty Acids Detection and Its Application in Taxonomy of Six Dictyostelid Cellular Slime Molds

Ying An, Pu Liu, Yu Li

Engineering Research Center of Chinese Ministry of Education for Edible and Medicinal Fungi, Jilin Agricultural University, Changchun, China

Abstract: Fatty acids of six species of Dictyostelid Cellular Slime Molds (Dictyostelids) were detected by using gas chromatography and the polymorphism was analyzed in order to investigate the relationship between the polymorphism of fatty acid and taxonomies of Dictyostelids in existence. Six species of Dictyostelids are *Polysphondylium candidum*, *Dictyostelium discoideum*, *D. implicatum*, *D. globisporum*, *D. clavatum*, and *D. tenue*. The fatty acids from these Dictyostelids were analyzed by cluster analysis. The results showed that the polymorphism of fatty acids of six species of Dictyostelids was evident based on fatty acid distribution. These Dictyostelids could be separated into two groups: group A and group B. Group A was *Polysphondylium candidum*, and group B were the other species of *Dictyostelium*. And 38 fatty acids from these Dictyostelids were also analyzed by cluster analysis; they could be separated into three groups: group I, group II and group III. Group I showed that frequency of occurrence of fatty acids was high, group II was in-between, and group III was low. Fatty acid might become a new index for taxonomy of Dictyostelids.

Key words: *Dictyostelium*; *polysphondylium*; cluster analysis; gas chromatogram

Foraging Behaviors of Phaneroplasmodia in Six Species of Myxomycetes to Three Types of Food Sources

Xiaoxia Song, Bao Qi, He Zhu, Qi Wang, Yu Li

Engineering Research Center of Chinese Ministry of Education for Edible and Medicinal Fungi, Jilin Agricultural University, Changchun, China

Abstract: Foraging behaviors of phaneroplasmodia of *Didymium megalosporum*, *D. squamulosum*, *Physarum melleum*, *Physarella oblonga*, *Badhamia gracillis* and *Fuligo septica* to three types of food sources (cereals, mushrooms and vegetables) were examined and compared under the same condition within 48 h. They had two foraging phases: explorative and exploitative growth reported in *P. polycephalum*. During explorative growth, their time of vigorous growth, migrating direction, contacting pattern and distribution mode were not related to new food sources. But, phaneroplasmodia of *D. megalosporum*, *D. squamulosum* and *P. melleum* were more vigorous than those of *P. oblonga*, *B. gracillis* and *F. septica*. During exploitative growth, their degree of biomass increase and resuming exploration were related to new food sources. Oat, *Lentinus edodes* and *Auricularia auricula* were favorable for their nutritional requirements, and especially *L. edodes* could be useful for their culture.

Key words: Physarales; foraging strategies; explorative growth; exploitative growth; nutritional status

Ultrastructure Observations on the Sporulation of *Physarum compressum*

Yanshuang Li, Xiaoli Wang, Yu Li

Jilin Agricultural University, Engineering Research Center of Chinese Ministry of Education for Edible and Medicinal Fungi, Changchun, China

Abstract: The plasmodia of *Physarum compressum* were cultured with sterile oat powder and water on the substrate of 2% water-agar, and the plasmodia turning to be sporangia at different stages to be surrounded by epoxide resin were selected, then cut into ultrathin sections, and observed by TEM (transmission electron microscope).

The observation results showed that the inner structures of proto-sporangia were denser than the plasmodia growing normally. The plasmodia were cleavaged into small spheres which were proto-spores. The proto-sporangia developed with plasmodia injection and the color changed from white to dark gradually. There was an electron-dense layer between two close spheres and two electron-lucent layers at the opposite sides of the electron-dense layer. When the color of sporangia getting black, we could see that the electron-dense layer became two layers, each layer becoming the outer spore wall with concave-convex parts which adapted to each other, and these were the ornamentations of the matured spores. And except for two layers, there was another thin electron-dense layer speculated as capllitia. So, some electron-dense layers didn't divide into two layers but three layers, and the middle layer would become capillitia after spores formed. It was extremely different from the traditional viewpoint that capillitium formed before spore wall. The results should be tested and verified by many other species, and then can be used in the taxonomy and systematics of Myxomycetes.

Key words: *Physarum compressum*; sporulation; ultrastructure

A Preliminary GC – MS Study of Four Species of Physarales

He Zhu, Qi Wang

Engineering Research Center of the Ministry of Education for Edible and Medicinal Fungi, Jilin Agricultural University, Changchun, China

Abstract: In this study, ultrasonic extraction and Saxhlet extraction of the fruiting bodies of four species of myxomycete and plasmodium of one species with petroleum ether were analyzed by gas chromatography-mass spectrometry (GC – MS). The extractions of liposoluble constituents were identified from the plasmodia of *Diderma chondrioderma* and the fruiting bodies of *D. chondrioderma*, *Diderma crustaceum*, *Craterium leucocephalum* and *Physarum cinereum*.

The results showed the compound structure patterns of the fruiting body and plasmodium of *D. chondrioderma* were mainly classified as aliphatic acids, aliphatic hydrocarbons, esters, higher aliphatic alcohol, higher aliphatic ketone and sterol, etc. The compounds of the fruiting body of *D. crustaceum* were mainly aliphatic acids, aliphatic hydrocarbons, esters, higher aliphatic alcohol, higher aliphatic aldehyde and steroids, etc. The compounds of fruiting body of *C. leucocephalum* were mainly aliphatic acids, aliphatic hydrocarbons, esters, higher aliphatic alcohol, etc. The compounds of fruiting body of *P. cinereum* were mainly aliphatic hydrocarbons, phenols, steroids, etc. Main chemical types of the compound were generally the same in different species of myxomycetes which belonged to the same order. The molecular structure of these compounds was relatively simple. Both the structure and type of aliphatic acids were typical aliphatic acids that once were found in marine algae and plankton. These aliphatic acids were particularly close to similar aliphatic acids found in amoeba, for example, 9 – hexadecenoic acid, n-hexadecanoic acid. This showed the evolution status of myxomycete to some extent and also confirmed the closer relationships between myxomycete and amoeba.

The analysis results of fruiting body of *D. chondrioderma* were compared with that of plasmodium. It showed that more than 50% chemical constituents of them were the same. This might be related to that they were the different life stages of the same myxomycete. It also showed that constituents of fruiting body were richer than plasmodium's from the spectrogram. This might mean a more complex physiological and biochemical process in the fruiting body.

The results showed there were several similar even the same compounds in the same species, genera and families respectively, such as high content of 5 - methyl - 2 - pyrrolidinone in the same order of Physarales. If it was a characteristic compound, we also need a lot of experimental data. Comparison of the analysis results of the spectrogram of *D. chondrioderma* with that of *D. crustaceum* and *C. leucocephalum*, showed that the proportion of the similar even the same compounds in the same genera is higher than that in different genera. The results showed that the species which had the similar genetic relationship had more similar GC - MS spectrogram, indicating their similar physiological and biochemical characteristics. So the chemical characteristics of myxomycete might be able to provide an important evidence of the study of genetic relationship of myxomycetes.

Key words: *D. chondrioderma*; *D. crustaceum*; *C. leucocephalum*; *P. cinereum*; fruiting body; plasmodium; GC - MS

Nuclear Observations of *Physarummelleum*

Qi Wang, Shu Li, Yu Li, Makoto Kakishima

Engineering Research Center of Chinese Ministry of Education for Edible and Medicinal Fungi, Jilin Agricultural University, Changchun, China

Abstract: Nuclear condition and its behavior are very important to understand life cycle of myxomycetes. Nuclear conditions of each step of life cycle were observed by Carl Zeiss Confocal Laser Scanning Microscope (LSM710) and its behavior was analyzed. Spores of *Physarum melleum* were used as materials as a first step of life cycle, and myxoamoebae and plasmodia were cultured on media. For staining nuclei DAPI was used and its fluorescence was observed with argon laser under the designated conditions. Fluorescence images were acquired in the sequential mode and processed using the software ZEN2012.

Key words: Nuclear; life cycle; confocal laser scanning; fluorescence

Liposoluble Constituents Comparison from Five Species of Myxomycetes

Wan Wang, Shu Li, Qi Wang

Engineering Research Center of Chinese Ministry of Education for Edible and Medicinal Fungi, Jilin Agricultural University, Changchun, China

Abstract: Myxomycetes is a group of eukaryotic organism with special life cycles and produces a number of metabolites with reported biological activities. In this study, we developed a method to process the sporophores of myxomycete before extracting with petroleum ether. Liposoluble constituents from sporophores of five species, *Diderma crustaceum*, *Lycogalaepidendrum*, *Stemonitis splendens*, *S. flavogenita* and *Arcyria obvelata* were extracted and analyzed by gas chromatography-mass spectrometry (GC – MS).

The results demonstrated that the molecular structure of most liposoluble constituents from the sporophores of species belonging to different orders was simple. Most of the liposoluble constituents were esters, phenols, aliphatic hydrocarbons, aliphatic acids, higher aliphatic alcohol, higher aliphatic aldehyde and aliphatic ketone and sterol. With close relationships among the species, *S. flavogenita* and *S. splendens* had similar compounds, such as alkanes, tetradecanal, oleic acids and unsaturated aliphatic acids. That also indicated the different liposoluble constituents among the orders. *A. obvelata* which belonged to Trichiales had a unique compound named ethyl citrate, which was not found in other samples belonging to Liceales, Physarales and Stemonitales.

In addition, more compounds of *S. splendens* were obtained by the modified method, such as oleic acid, stigmasterol. It indicated that the new method could increase the efficiency of spore wall breaking and petroleum ether extraction.

Key words: Slime mould; sporophore; liposoluble constituent

Description of the Amoeboid Movement of Myxamoebae in Several Myxomycetes Species

Xiaoli Wang[1,2], Chen Li[2], Yu Li[1]

1. Engineering Research Center of Chinese Ministry of Education for Edible and Medicinal Fungi, Jilin Agricultural University, Changchun, Jilin, China;
2. College of Agronomy, Jilin Agricultural University, Changchun, Jilin, China

Abstract: In nature, myxomycete spores probably germinate and issue one or more myxamoebae. Myxamoebae is different from diverse slime molds, and the main difference is the presence or absence of transparent ectoplasm and regularity or irregularity of the shape. Transparent ectoplasm may be responsible for the pattern of amoeboid movement. It is difficult to form typical pseudopodium for myxamoeba with thick ectoplasm, and exhibits gliding motility with slow speed. The others with less ectoplasm will form lobopodium or filopodium and exhibit flowing motility and creeping motility. In order to describe accurately the motility pattern of myxamoebae, in this research, the myxamoebae of *Hemitrichia calyculata* and *H. clavata* (Trichiales), *Physarum melleum* and *P. globuliferum* (Physarales), and *Stemonitis flavogenita* (Stemonitales) was used as the experimental material, with the help of graphing software, to paint trajectory and calculate movement velocity, draw the instantaneous perimeter of myxamoeba which show the change speed of pseudopodia. All above can visualize the movement characteristics of myxamoeba.

Key words: Myxamoeba; transparent ectoplasm; pseudopodium; amoeboid movement

Species Diversity of Myxomycetes on Different Decay Stages of Coarse Woody Debris in Laurel Forest of Warm Temperate Western Japan

Yuichi Harakon, Shoji Ohga

1. Asakita-Highschool, Miirihigashi, 1 – 14 – 1, Asakitaku, Hiroshima, 731 – 0212, Japan;
2. Department of Forest Product Science, Kyushu University, Tsubakuro 394, Sasaguri, Kasuya, Fukuoka 811 – 2415, Japan

Abstract: The distribution of myxomycete species on coarse woody debris (CWD) associating with decay state was little investigated in laurel forests of warm-temperate ecosystem in the world. The present study carried out entire year surveys during 2006 – 2008 and revealed 70 species on decayed evergreen logs from totally 1079 samples. Myxomycetes occurred on various decaying stages of wood which contained sufficient moisture. Most species of 81% in total species occurred on moderately decayed wood with the highest species diversity. Dominant species were four species, i. e. *Ceratiomyxa fruticulosa* (O. F. Müll.) T. Macbr. , *Lycogala epidendrum* (L.) Fr. , *Physarum viride* (Bull.) Pers. and *Hemitrichia calyculata* (Speg.) M. L. Farr. The 42 species recorded with eight or more samples were arranged in order of succession index corresponding to the stage of decay. Species of Physarales characteristically dominated on hard wood, while species of Trichiales did on decayed softer wood. According to decay stage dominant species were distinctive, most species had a preference to moderately decay stage of wood, while several species occurred on hard and (or) softer decayed wood. *Physarum viride* (Bull.) Pers. occurred on hard wood, *C. fruticulosa* on moderately decayed wood, *L. epidendrum*, *Hemitrichia calyculata* (Speg.) M. L. Farr, *Arcyria denudata* (L.) Wettst. , and *Arcyria cinerea* (Bull.) Pers. on decayed softer wood, and *Cribraria* tenella Schrad. on brittle decayed wood. Consequently the myxomycete assemblages were the lowest similarity between the different stages of decaying progression, hard wood and brittle decayed wood. It was quantitatively revealed that the myxomycetes intently inhabit with the highest diversity on the moderately decayed CWD of the angiospermous evergreen trees.

Key words: Moisture; dominant species; Physarales

后 记

科学技术是第一生产力。纵观历史,人类文明的每一次进步都是由重大科学发现和技术革命所引领和支撑的。进入 21 世纪,科学技术日益成为经济社会发展的主要驱动力。我们国家的发展必须以科学发展为主题,以加快转变经济发展方式为主线。而实现科学发展、加快转变经济发展方式,最根本的是要依靠科技的力量,最关键的是要大幅提高自主创新能力。党的十八大报告特别强调,科技创新是提高社会生产力和综合国力的重要支撑,必须摆在国家发展全局的核心位置,提出了实施"创新驱动发展战略"。

面对未来发展之重任,中国工程院将进一步加强国家工程科技思想库的建设,充分发挥院士和优秀专家的集体智慧,以前瞻性、战略性、宏观性思维开展学术交流与研讨,为国家战略决策提供科学思想和系统方案,以科学咨询支持科学决策,以科学决策引领科学发展。

工程院历来重视对前沿热点问题的研究及其与工程实践应用的结合。自 2000 年元月,中国工程院创办了中国工程科技论坛,旨在搭建学术性交流平台,组织院士专家就工程科技领域的热点、难点、重点问题聚而论道。十年来,中国工程科技论坛以灵活多样的组织形式、和谐宽松的学术氛围,打造了一个百花齐放、百家争鸣的学术交流平台,在活跃学术思想、引领学科发展、服务科学决策等方面发挥着积极作用。

中国工程科技论坛已成为中国工程院及至中国工程科技界的品牌学术活动。中国工程院学术与出版委员会将论坛有关报告汇编成书陆续出版,愿以此为实现美丽中国的永续发展贡献出自己的力量。

中国工程院

郑重声明

高等教育出版社依法对本书享有专有出版权。任何未经许可的复制、销售行为均违反《中华人民共和国著作权法》，其行为人将承担相应的民事责任和行政责任；构成犯罪的，将被依法追究刑事责任。为了维护市场秩序，保护读者的合法权益，避免读者误用盗版书造成不良后果，我社将配合行政执法部门和司法机关对违法犯罪的单位和个人进行严厉打击。社会各界人士如发现上述侵权行为，希望及时举报，本社将奖励举报有功人员。

反盗版举报电话　（010）58581897　58582371　58581879
反盗版举报传真　（010）82086060
反盗版举报邮箱　dd@hep.com.cn
通信地址　北京市西城区德外大街4号　高等教育出版社法务部
邮政编码　100120